CULTURES OF MASS TOURISM

New Directions in Tourism Analysis

Series Editor: Dimitri Ioannides, Missouri State University, USA

Although tourism is becoming increasingly popular as both a taught subject and an area for empirical investigation, the theoretical underpinnings of many approaches have tended to be eclectic and somewhat underdeveloped. However, recent developments indicate that the field of tourism studies is beginning to develop in a more theoretically informed manner, but this has not yet been matched by current publications.

The aim of this series is to fill this gap with high quality monographs or edited collections that seek to develop tourism analysis at both theoretical and substantive levels using approaches which are broadly derived from allied social science disciplines such as Sociology, Social Anthropology, Human and Social Geography, and Cultural Studies. As tourism studies covers a wide range of activities and sub fields, certain areas such as Hospitality Management and Business, which are already well provided for, would be excluded. The series will therefore fill a gap in the current overall pattern of publication.

Suggested themes to be covered by the series, either singly or in combination, include—consumption; cultural change; development; gender; globalisation; political economy; social theory; sustainability.

Also in the series

The Framed World
Tourism, Tourists and Photography
Edited by Mike Robinson and David Picard
ISBN 978-0-7546-7368-2

Brand New Ireland?
Tourism, Development and National Identity in the Irish Republic
Michael Clancy
ISBN 978-0-7546-7631-7

Cultural Tourism and Sustainable Local Development
Edited by Luigi Fusco Girard and Peter Nijkamp
ISBN 978-0-7546-7391-0

Crisis Management in the Tourist Industry
Beating the Odds?
Edited by Christof Pforr and Peter Hosie
ISBN 978-0-7546-7380-4

Cultures of Mass Tourism
Doing the Mediterranean in the Age of Banal Mobilities

Edited by

PAU OBRADOR PONS
University of Sunderland, UK

MIKE CRANG
Durham University, UK

PENNY TRAVLOU
Edinburgh College of Art, UK

ASHGATE

Published by
Ashgate Publishing Limited
Wey Court East
Union Road
Farnham
Surrey, GU9 7PT
England

Ashgate Publishing Company
Suite 420
101 Cherry Street
Burlington
VT 05401-4405
USA

www.ashgate.com

British Library Cataloguing in Publication Data
Cultures of mass tourism : doing the Mediterranean in the
    age of banal mobilities. -- (New directions in tourism
    analysis)
    1. Tourism--Social aspects--Mediterranean Region.
    2. Culture and tourism--Mediterranean Region.
    I. Series II. Obrador-Pons, Pau. III. Crang, Mike.
    IV. Travlou, Penny.
    306.4'819'091822-dc22

Library of Congress Cataloging-in-Publication Data
Pons, Pau Obrador.
    Cultures of mass tourism : doing the Mediterranean in the age of banal mobilities / by Pau Obrador Pons, Mike Crang and Penny Travlou.
        p. cm.
    Includes index.
    ISBN 978-0-7546-7213-5 (hardback) 1. Tourism--Mediterranean Region--Marketing.
2. Culture and tourism--Mediterranean Region. 3. Package tours--Mediterranean Region. 4. Mediterranean Region--Social conditions. 5. Mediterranean Region--Economic conditions. I. Crang, Mike. II. Travlou, Penny. III. Title.
    G155.M46P66 2009
    306.4'819091822--dc22

2009016975

ISBN 9780754672135 (hbk)
ISBN 9780754697763 (ebk)

Mixed Sources
Product group from well-managed
forests and other controlled sources
www.fsc.org Cert no. SA-COC-1565
© 1996 Forest Stewardship Council
FSC

Printed and bound in Great Britain by
MPG Books Group, UK

Contents

List of Figures

List of Authors

Rachele Borghi is a Research Fellow at the University of Venice

Javier Caletrío is a Research Fellow at the Centre for Mobilities Research at Lancaster University.

Mike Crang is a Reader in Human Geography at Durham University, UK.

Michael Haldrup is Associate Professor in the Department of Geography and International Development Studies, Roskilde University, Denmark

Claudio Minca is Professor of Geography at Royal Holloway, the University of London and Visiting Professor at the College of Tourism, Rikkyo University, Tokyo.

Dan Knox is a Senior Lecturer in Tourism at the Centre for Tourism, Consumer and Food Studies, Liverpool John Moores Univeristy, UK.

Pau Obrador Pons is a Lecturer in Tourism, Faculty of Business and Law, University of Sunderland, UK.

Karen O'Reilly is a Reader in Sociology at the University of Loughborough, UK.

Penny Travlou is a Research Fellow at the OPENspace Research Centre, Edinburgh College of Art/Heriot-Watt University, UK and Lecturer in Cultural Geography at the Centre for Visual and Cultural Studies, Edinburgh College of Art.

Chapter 1

Introduction: Taking Mediterranean Tourists Seriously

Pau Obrador Pons, Mike Crang and Penny Travlou

This book is about the emerging cultures of mass tourism. While the rise of mass tourism has been the subject of much attention, the focus has tended to be on the impact it has upon local peoples, local economies or local environments as well as on economic and managerial issues. This book instead wants to take the cultures of mass tourism seriously, knowing that even putting it so sounds like an oxymoron to many. With more than 230 million international tourists a year, the Mediterranean is the largest tourist destination region in the world. To follow Löfgren (1999) the Mediterranean is now less united by Braudel's rhythms of olive, grain and wine cultivation, shared trade and Roman legacies than the fortnightly pulse of the package tour, the circulation of resort types and the shared culture of sun-seeking tourism. This book sets out to demonstrate that the economic importance of mass tourism in the Mediterranean is matched by its significance as cultural and aesthetic phenomena. As well as transforming the derelict economies of the Mediterranean, mass tourism is one of the most sensational cultural phenomena of our times and is a central feature of the contemporary everyday in Western Europe (Rojek 1993, Inglis 2000, Franklin and Crang 2001). It can no longer be considered as a discrete activity, contained in special locations and times, but paraphrasing Franklin it 'has become a metaphor for the way we lead our everyday lives in a consumer society' (2003: 5). Mass tourism has developed into a new cultural formation that mixes global, national and local influences. Most of the elements of this cultural formation, which have yet to be addressed, destabilize fixed and coherent identities for places. Mass Mediterranean tourism is rather creating a new space of related and refracted practices through reciprocally entwined, though not equal, cultures of work and tourism. It is these new practices and spaces that this collection brings together.

Mass Mediterranean tourism is a historically specific phenomenon that is generally associated with three different elements. First and foremost, it is associated with the democratization of leisure and the extension of tourism to all sectors of society. It is in this respect a 'quantitative notion' that refers to the 'proportion of the population participating in tourism or on the volume of tourist activity' (Bramwell 2004: 7). Secondly mass tourism is also associated with a particular mode of tourism production that emphasizes economies of scale. Mass tourism involves the industrialization of leisure, the translation of Fordist principles of accumulation to tourism, including the large replication of standardized products, the reduction of costs and the promotion of mass consumption and spatial and temporal concentration. Cheap package holidays are the most visible manifestation of this mode of production, which is currently being replaced by more sophisticated versions combining economies of scope with economies of scale. Finally, mass tourism is associated with a particular tourist sensibility that emphasizes warm climate, coastal pleasures, freedom from the regulated world, relaxation and a party atmosphere. Often defined in opposition to classical ideas of travel as sightseeing, mass tourism represents a different tourist experience that is summarized with the three 'S's: sun, sea and sand (and also sex and spirits).

These three elements come together for the first time in the late nineteenth century in the coastal towns of Britain, notably Blackpool (Bennett 1983, Walton 2000). After the Second World War mass tourism internationalizes, establishing the Mediterranean as its main destination and its most remarkable manifestation. A number of factors made possible the emergence of mass tourism including innovations in transport, the consolidation of a welfare system, the increase in real income, the reorganization and rationalization of time and the improvement of international relations. Over the past 20 years or so mass tourism has embarked on a still ongoing process of restructuration that has profoundly reshaped the neat picture that emerged in the 1950s and 1960s (Bramwell, 2003, Aguiló 2005). The introduction of more flexible forms of accumulation in tourism has lead to the expansion of mass tourism beyond the beach, with a diversification of tourist experience though the industrialization of leisure as well as climatic and hedonistic pleasures remain a common denominator in the majority of Mediterranean tourism. In this context it would be a mistake to reduce mass tourism to beach tourism and indeed to ignore its relationship with everyday life in Northern Europe.

Mass tourist sites are some of the most iconic examples of western consumer societies; and yet the study of tourism is still dominated by policy led work, with a productivist bias that emphasizes the topics and values linked to processes of production. Issues related with the organization of tourist provision and its impact have systematically received priority over issues related with touristic consumption. Even when consumption is addressed it tends to be through the production of marketing materials, and the creation of landscapes for consumption, turning mass tourism into 'a logical extension of the general principle of industrial capitalism to the realm of leisure' (Böröcz in Koshar 1998: 325). The majority of literature on mass tourism and beach resort tourism, including the works of Bramwell (2004), Apostolopoulos et al. (2000) and Montanari and Williams (1995), adopts an economic and managerial perspective on resort development, growth and increasingly, sustainability. Some of these works make reference to cultural change, focusing mainly on the impacts of mass tourism on the host community. However studies of the culture of tourists are relatively uncommon and those of the culture of mass tourism even rarer. Simply taking the holdings of one of the editor's university catalogues, less than 10 percent of the more than 200 volumes on tourism were significantly focused around the tourists themselves, their experiences and what they made of them And of those, most were then picking out specific interesting niches of cultural tourists (such as backpackers or spiritual tourists). This seems to reflect the origins of much of the interesting work on cultures of tourism through anthropology. There the tradition of the study of isolated local communities has often set up an uneasy relationship with tourism and tourists, where their presence is seen as a threat to local identity (and the credential of professional travellers and scholars) (Crick 1992, Risse 1998, Galani-Moutafi 2000, Crang 2009). The effect has been that cultural studies of tourism are freighted with problematic legacies of scholarship that saw 'host' communities as static, small scale recipients of industrially produced flows of people. Local cultures are seen as eroded by a homogenous inauthentic, consumer culture.

The study of mass tourism is shaped through a series of striking paradoxes. For sure, mass tourism does promise pure entertainment and often flirts with the banal; and yet the conceptualization of tourism still relies on outdated notions of authenticity that establish social distinctions between good and bad tourism. Much critical academic work focuses upon breaking down those promises to reveal the often unpalatable social realities involved in their production. Inglis (2000: 5) notes 'The dreams are powerful and beautiful' and we should be careful of dismissing them and their role in people's lives, even if 'dedicated dreambusters in their big boots will, correctly, point out the horrors and boredom of actually existing tightly packaged trips, the mutual exploitation of tourist and native'. Mass tourism is thin on meaning and ideological narratives and very dense on physicality and sensuality; and yet scholars with an interest in cultures of tourism have chosen overwhelmingly to examine discourses, meanings and ideological structures at the cost of physicality. With the emergence of mass tourism, the Mediterranean has been re-integrated within a global set of cultural, social and economic networks; however the Mediterranean is still conceptualized as a bounded region that is subject of external forces producing impacts, a region that needs to be preserved from foreign invaders. This book responds to our discomfort with these shortcomings. Dominant perspectives on tourism have failed to provide an adequate basis for exploring the cultural dimension of mass tourism.

Despite its enormous significance mass tourism rarely makes an appearance in contemporary social science, which tends to marginalize this tourist phenomenon. With few exceptions, most notably Urry (1990) and Bauman (2000), tourism is a stranger to current theorizations on consumption, globalization, identity formation and the consolidation of new modes of bio-political regulation. If used as an example, it is often of a shallow identity or subjectivity. In recent years there have appeared several high profile collections and volumes reflecting a surge in interest in cultures of tourism and their significance. The most significant of these are Urry and Sheller's *Tourism Mobilities* (2004), Desmond's (1999) work on *Staging Tourism*, the histories of vacationing by Löfgren (1999) and Inglis (2000), the collections by Crouch on leisure and tourism geographies (1999), by Bærenholdt et al. on 'Performing Tourist Places' (2004) and by Minca and Oakes on the paradoxical desires and outcomes of tourism (2006). All these works interrogate the practices and cultures of tourism as this volume does, but the beach and mass tourism form only a fairly limited aspect of the whole. Most of the works on cultures and practices of tourist still focus on specialist forms of usually high status tourism, such as ecotourism, literary and heritage, adventure tourism. There are, nevertheless several great works looking at 'living with' or 'coping with' mass tourism in the Mediterranean (Boissevain 1996, Waldren 1996, Abram et al. 1997) and tourist representations (Selwyn 1996). There are also some interesting examples of work devoted to the sun and sand tourism and cultures of the beach (Lenček and Bosker 1998, Urbain 2003), the lure of the sea (Corbin 1994) and sunlight (Carter, 2007). Despite all these valuable contributions, the prevailing view is still that exemplified in the classic work of Turner and Ash (1975) which vilifies mass tourism as uncultured, uncaring and alienating. They describe mass tourists as 'the barbarians of the age of leisure, the golden hordes' and the Mediterranean as 'a pleasure periphery' (1975: 1). Confronted with the all too apparent constrictions and obvious exploitations of mass tourism, tourist studies generally downplays the banal, the un-exotic and, in particular, the pleasurable character of the tourist experience, reproducing the binary opposition between travel and tourism. It is these moments of actual existing pleasure that this work seeks to recover and to which it gives serious attention, balancing the horrors and boredom with the dreams and hopes, the exploitations with the liberties (Inglis 2000: 5). 'It is important', Löfgren reminds us, 'to see that standardized marketing does not have to standardize tourists. Studies of staging of tourist experience in mass tourism often reduce or overlook the uniqueness of all personal travel experience' (1999: 8). The lack of attention to the cultures of mass tourism, especially the dreamings and doings of mass tourists, is a major shortcoming in tourist studies. Dominant frameworks – heavily dependent on romantic ideas of travel – dismiss at the best of times the liveliness and creativity of mass tourism. More often than not they fail to unpack the phenomenon at all.

This edited collection contributes to the study of mass tourism with a series of ethnographical insights into some of the key sites of this tourist phenomenon, including the villa, the beach, the island and coastal hotel. In so doing, we want to extend the surge of interest in cultures and practices of tourism beyond specialist products, bringing to light a major component of contemporary consumer culture in Western Europe. The prime focus of this book is on the mundane and banal aspects of mass tourism. We argue for attentiveness to the diversity of practices in mass tourism and look at theorizing it as a way of being in the world, as materially constituted and constituting a social world, being alongside other people and a way of relating to places. The book has an unequivocal empirical orientation with all chapters reporting from recent and ongoing field research in the Mediterranean. However underlying the book there is also a deep rooted theoretical concern with developing new perspectives on mass tourism, new ways of looking at and thinking of this tourist phenomenon that break with the shortcomings of dominant perspectives. Equally we believe it is important to break with the tendency to isolate this tourist phenomenon from the main debates in the social sciences. The field of mass tourism has an enormous potential to be a fertile ground for developing social theory, in particular that relating to contemporary consumer culture. In this introductory chapter we summarize some of the theoretical concerns underpinning this book. We identify three broad areas of inquiry relating with banality and biopolitics, the spatial and temporal dimension of mass tourism and its enactment. It is not our intention to set out a canonical perspective on mass tourism, but to identify some emerging research agendas. Our intention is to contribute to the renewal of a field in desperate need of fresh ideas.

Banality and Biopolitics on the Beach

Rather than adopting an economic or managerial perspective, this book sets out to demonstrate the cultural significance of mass tourism and the significance of mass tourism to mass culture. Such an approach encounters its most serious challenge in the perceived banality of mass tourism. We are confronted with tourist phenomena that draw on gritty vulgarity, playful crowds, a culture of indulgence, a series of corporeal pleasures and the blend of the ancient with the ironic and the kitsch among other things. Mass tourism offers a distinctive form of entertainment – more 'vulgar' and 'corporeal' – that clashes with the sophistication and detachment of middle class forms of travel, the values of which underpin dominant conceptualizations of tourism. The significance of the banal in tourism has been systematically overlooked by dominant perspectives which have privileged the exotic and the spectacular. There has been little interest and respect for the banal practices and pleasures of ordinary tourists. In downplaying the banal, dominant perspectives have reproduced a social hierarchy of travellers and tourists, thus sanctioning a set of ideological and social distinctions that is as much a stake in class distinction as an actual description of tourist practices. As Franklin and Crang point out 'too often we risk treating the numerous and enumerated tourists as foreign species, "Turistas Vulgaris", only found in herds, droves, swarms and flocks' (2001: 8). Mass tourism might be a 'depthless' and a fluid phenomenon with few meanings and utopias attached but it still is a site of relevant social and cultural practices that speak mainly to the body and the sentient. A cultural perspective on mass Mediterranean tourism demands reflection on these elements, including the ordinary experiences of the package tour, the proliferation of highly commodified environments devoted to leisure, the hedonism of night life in Eivissa and Faliraki and the corporeality of sunbathing, among others.

How to make sense of the banal in tourism is one of the main concerns of this book. Drawing on the Frankfurt School some scholars prefer to proclaim the insubstantiality of the banal spaces of late capitalism, emphasizing their impoverishing effects on social life. Others working within the theoretical framework of identity politics prefer the celebration of the banal as a site of political resistance, taking it as a symbolic expression of the more disadvantaged sectors of society. As Meghan Morris noted twenty years ago 'banality' serves as a mythic signifier that has all too easily served as a mask for the question of value, and of 'discrimination' (Morris 1988: 27). Instead, the route this edited collection follows emphasizes the multiple moods and mutations of the practical, the ordinary and the everyday, that is the banal (Billig 1995). Tourism is part of a 'banal seduction' then, not some Baudrillardian 'fatal shore' of catastrophic cultural negation, but the ongoing enchantment yet circumscription of life. We start from Haldrup's, 'banalizing tourism', where the time–space of tourism and the everyday permeate each other, but this book goes further in calling attention to the banal itself as being the ways tourist practices produce and reproduce social life and materialize structures of feeling and moral dilemmas. We suggest it is 'enabling' for tourist studies and not merely 'something that is left behind after it has been exorcized or redeemed in the movements of cultural analysis' (Seigworth 2000: 229). Tourism does not need to be recovered from banality, either by finding the exceptional within it nor by finding its normality exceptional. Banal desire is grounds for neither condemning tourism as beyond redemption nor redeeming some putative resistance – it is instead the medium of tourism. Thus paying attention to it is to heed what Michel de Certeau called 'the oceanic rumble of the ordinary' where

> the task consists not in substituting a representation for the ordinary or covering it up with mere words, but in showing how it introduces itself into our techniques – in the way in which the sea flows back into pockets and crevices in beaches – and how it can reorganize the place from which discourse is produced (de Certeau 1984: 5)

This collection thus faces the difficulty of much work on mass culture in avoiding ventriloquizing the ordinary, or creating some monolithic, undifferentiated everyday sense that is the same for everyone and applicable to no one, or romanticizing the popular to invest the study with glow of resistance. With different degrees of intensity the contributors to this book seek to address this by dislocating attention away from symbolic meanings and discourses to the actual everyday doings and enactments of tourists and their role in the production of social meanings and knowledges. Turning attention towards embodied social performances opens the possibility to acknowledge the significance of the banal in tourism and escape the conceptual straight-jacket that has prevented social science making sense of mass tourism.

The route we follow involves recovering a sense of culture beyond traditional elitist or exotic forms to include the ordinary experiences of common people, but without seeing those terms as defining each other. It also involves the stretching of culture beyond the rational and the visible to include the everyday invisible elements that make up social life, the lay and popular knowledges, the habits, skills and conventions as well as the unreflexive practices. Culture here is not a fixed, finished product but a historically specific formation that has to be continuously enacted. As Edensor (2001) explains, tourism is a mundane system of practice and performance, a highly regulated and choreographed space as well as a realm of improvization and contestation. In developing a pragmatic sense of mass tourism, this approach is careful not to reproduce social distinctions between travel and tourism, seeking instead to 'come to terms with the continual oscillation around the poles of traveller and tourist' (Franklin and Crang 2001: 8). Rather than bringing together all sorts of travel within a single hierarchy, this approach seeks to reconstruct the heterogeneous histories and trajectories of tourism. An emphasis on the ordinary and the everyday requires a higher attention to the diversity of practices in tourism.

It would be wrong to assume – as many scholars implicitly do – the innocence and simplicity of banal tourist practices. As Billig explains, 'banal does not mean benign' (1995: 4). Still marked with meaning, banal forms of tourism have proved to be an effective vehicle for the (re)production of social and cultural formations, making them look normal and ordinary while placing them out of public scrutiny. Mediterranean tourism has played an important geopolitical and ideological role (Pack 2006); however the political and cultural significance of mass tourism is first and foremost at the level of the body and the sentient. The Mediterranean is a major cultural laboratory for the production of bodies, feelings and subjectivities in Western Europe. The most critical role mass tourism has played relates the formation of postmodern consumer sensibilities. The notion of bio-politics (Foucault 1978) provides a useful framework to re-position mass tourism in relation to contemporary cultural processes. Mass tourism is a prominent example of a new form of biopolitical circulation that consolidates in Western Europe after the Second World War with the emergence of post-industrial globalized economies increasingly reliant on leisure and consumption (Vilarós 2005). Drawing on Foucault, bio-politics is understood here as the 'Extensive complex of discourses, practices and institutions tasked with the care, regulation, and improvement of individual bodies and of the collective body of national population' (Vasudevan, 2006: 800). This new form of bio-political circulation is more interested in the production of life than the power over life. Mass tourism emerges in a transitional context from primary systems of power based on technologies of repression towards a much more vitalist and creative political systems which pursue the incorporation of citizens within a new order of hedonism and consumption. Mass tourism points to the importance of play (Huizinga 1998), the carnivalesque and grotesque (Bakhtin 1984) in contemporary culture. If the celebration of *homo ludens* is one part of a globalized hedonism, then mass tourism has also long played upon the body beautiful and the bodily grotesque in such things as the British tradition of sexualized yet ironic seaside postcards (Löfgren 1985).

The carnivalesque also points us to the phenomena of the crowd, the golden horde of Turner and Ash – whose numerousness and density challenges notions of romantic contemplation. Le Bon (1895/2002) figured the twentieth century as the age of the crowd, and one where the crowd were the barbarians breaking down social orders. A more vitalist approach renews our attention to the sociality of crowds, as more than individual actors but ones where writers like Gabriel

Tarde emphasized the plastic qualities of individuation and the trans-subjective dimensions of experience. The crowd for him is a mix of spontaneity and somnambulism. This is a tradition that has been eclipsed by later sociological studies that have tended to focus upon the aggregation of individuals rather than the emergent properties of spontaneous generation of imitation (Borch 2006). This surely gives us a sense of the creative regulation of the sociality of a destination like Benidorm. Here we might see the emergence of Bacchanalian partying in 'a space that has become the most effective substitute for the time of the breaking-up party, that countryside festival that industrialization eliminated from the calendar of Europeans' (MVRDV 2000: 105). Not only are the beaches densely populated but tourists spend almost as long wandering the streets (up 3.25 hrs per day in town, 3.5 hrs on the beach) so that 'tourists surveyed in Benidorm are notable for their passion for the streets… incited by its charms, which they find unending, moved by surprise, encouraged by he possibility of meeting and recognizing and urged on by the fleeting nature of their stays '(MVRDV 2000: 112). This is a biopolitics of density and proximity producing an affective charge. In short we need to take seriously the emotional productivity of the *mass* in mass tourism and to actually think about how a 'mass' functions.

Although very few scholars with an interest in tourism turn to the concept of bio-politics, a concern for the production and regulation of bodies, affects and subjectivities is major theme in cultural accounts of Mediterranean tourism. In a ground-breaking history of the holiday, Inglis (2000) emphasizes the role of the Mediterranean in the creation of new kinds of feelings. Carter's (2007) original cultural account of sunlight also identifies the Mediterranean as one of the main stages in the constitution of the contemporary, sometimes problematic relationship with the sun. Similar logics can be found in the history of vacationing of Löfgren (1999), the work of Urbain (2003) on the beach, or in MVRDV's analysis of the new urban landscapes emerging in the Iberian coast, where they see a linear city ringing Iberia with a 'wall of banality' (MVRDV 2000: 75). Despite their differences, all of these accounts point to the significance of banal spaces and practices of mass tourism in the production of new postmodern forms of life. The contributors to this book also share the belief that mass *banal* tourism is not a social and cultural desert. The banal is central to most chapters of the book: Minca and Borghi explore the translation of a new Orientalist images of Moroccan into a series of mundane tourist practices and landscape aesthetics; Haldrup looks at the significance of tourism as a banal form of experiencing orientalism and cosmopolitanism; Obrador Pons develops insights into the mundane ways spaces of hospitality are inhabited and Caletrío emphasizes the quotidian and domestic rituals of tourist life in Costa Blanca. Ultimately this book emphasizes the need of a more creative approach to mass tourism that neither celebrates nor dismisses the '*banal*', the '*corporeal*' and the '*vulgar*'; an approach that focuses on the productive effects of mass tourism rather than on its ideological underpinnings. It is an approach that sees tourism as an abstract machine rather than an abstract rationality (Guattari 1992). A machine that combines economic, technological, symbolic, emotive and bodily registers to produce and regulate spaces, affects and practices.

The Time and Space of Mass Tourism

If mass tourism is seen as being tainted by the banal then it is in turn seen as degrading and defiling the places with which it comes into contact. This vision is neatly summed up by Guy Debord in his *Society of the Spectacle*

> Capitalist production has unified space, which is no longer bounded by external societies. This unification is at the same time an extensive and intensive process of *banalization*. The accumulation of commodities produced in mass for the abstract space of the market, which has … destroyed the autonomy and quality of places. (Debord 1973: 165)

Here then the banal becomes linked with the erosion of authentic place, and tourism becomes the literal personification of commodification:

> Tourism, human circulation considered as consumption, a by-product of the circulation of commodities, is fundamentally nothing more than the leisure of going to see what has become banal. The economic organization of visits to different places is already in itself the guarantee of their equivalence. The same modernization that removed time from the voyage also removed from it the reality of space. (Debord 1973: 168)

This powerful denunciation looks at tourism as producing a form of space but an abstract one of equivalences. Mass tourism is often offered as the example *par excellence* of what Augé (1995) calls non-place, with its lack of local references in resorts, beyond the usual suspects of sangria and paella, or ouzo and moussaka. Debord's hyperbolic vision has echoes in many tourism studies that rely not only on the problematic sense of banality but also take for granted an organic notion of authentic place. Organicist perspectives on place draw on a nostalgic vision of a stable and harmonious rural locale, in which there is a coincidence between place and community and between dwelling and sedentarism. What gives a place its specificity and identity is endurance, self-containment and inward-looking history. In this approach proximity and isolation make places whereas mobility and diversity threatens their existence. Stable rural communities based on face-to-face relationships tend to be identified as genuine and authentic whereas the urban, mobile and fragmentary life of modernity tends to be seen as alienating and destructive. This opposition between organic communities and the world of flows of globalization and tourism has been reworked through a number of binaries such as local versus global, authenticity versus commodification, traditional versus modernity and indeed cultural versus economic (Crang 2006: 55). These binary oppositions inevitably feed back into policy analysis, with a common distinction between good and bad tourism, the latter of course is always mass tourism. The tendency to define tourist places as bounded and enclosed has been denounced in recent years (Crang 2006, Minca and Oakes 2006). As Crang points out tourist studies 'produce an oddly fixed version of the world for a mobile and fluid process' (2006: 48). The main problem with linking authentic dwelling and fixity is that there is no possibility of authenticity and dwelling in the hypermobile spaces of mass tourism, which instead become the repository of all sorts of spatial illnesses including commodification, displacement and alienation.

We believe an ontological shift in the study of mass tourism is necessary that takes into account the different mobilities underpinning this tourist phenomenon. If we are to demonstrate the cultural significance of mass tourism, we need to mobilize both tourists and places in our analysis, breaking with theoretical perspectives that deny the possibility of dwelling-in-mobilities, that is, that deny the place-making capacity of fluid social formations such as mass tourism. We are not suggesting abandoning place in favour of movement but reconceptualizing place so that it incorporates the possibility of movement and displacement rather than seeing them as antithetical terms. Tourism places are constituted through many forms of mobilizations and demobilizations among people (both tourists, workers and 'locals'), images and things, at different scales and speeds, where some mobilities are predicated on restricting others, and senses of relative fixity and transience, and some people and things re remobilized in new ways (Sheller 2004: 15, Crang 2009). In other words, these more or less transient phenomena come together in conjunctions that create the tourist landscape. The study of mass tourism needs to embrace a relational approach to place (Massey 2005). Places are not fixed and stable entities but they are provisional and always in the process of construction. Their shape depends on the performances and interactions of the people that inhabit them and the networks that sustain them. As Urry and Sheller point out 'Places are about relationships, about the placing of peoples, materials, images and the systems of difference they perform' (2004: 6). The extraordinary fluidity of contemporary mass tourism is generative of unique spatialities (and temporalities), which are central to the constitution of postmodern subjectivities.

Mass tourism is no longer confined simply to the spatiality of the resort nor to the rhythms of the fortnight vacation, in either its operations or its effects. Even exceptional tourist localities such as Magaluf and Benidorm are shot through with banal connections to the rest of life. Multiple unrelated geographies and temporalities intervene in the constitution of this tourist phenomenon, in a continuous game of circulation and emplacement, of deterritorialization and re-territorialization. Mass tourism may come to light in the beaches of Benidorm, but it is already present in the cold dark winter days of the industrial cities of Germany and the United Kingdom to which it will return when the summer is over in the form of postcards or Ibiza Mix compilations of techno music. It is precisely this capacity of going back and forth and passing through and dwelling in between that confers uniqueness to the objects and spaces of mass tourism. Paraphrasing

Crang, mass tourism 'aims to produce [...] a sense of being somewhere different and specific, but it does so through a number of constitutive absences' (2006: 63). We need to destabilize our objects of study from the destination and resort, that comprise so much tourism analysis. Being a tourist is not a solid and permanent state of mind, but a temporary subjectivity that is fragile and elusive. It is a rhythmical phenomenon that appears and disappears. In the summer months the Mediterranean beaches may be full of bodies but a few months later there are only ghostly reminders of them – traces of bodies on beaches and of beaches on bodies. Very few pieces of work have been able to capture this sense of dislocation and fluidity. Moreover highlighting the fluid nature of doing tourism destabilizes the tired opposition of mass tourist and niche tourist, as cultural tourists use mass facilities, as weekend breaks become commonplace over greater distances, when tourists become residents or locals become returning émigrés. These ambiguities are highlighted in the work of Karen O'Reilly on the Costa del Sol (2000) which she develops in this book (Chapter 7). Michael Haldrup in Chapter 3 develops a similar strategy, that dislocates the objects of study. This book does not look then at the impact of tourist globalization on local cultures but the emergence of new transnational cultures which involve different levels of presence and absence, local and global relations for all.

As well as rescaling tourism, taking into account the different mobilities and displacements that constitute it, it is also necessary to situate this tourist phenomenon within specific historical and social contexts. It is commonly believed that mass tourism is an external phenomena imposed by the forces of globalization that bares no relation with the identity of the Mediterranean, a manifestation of the economic and social dependency of the south in relation to the north. We would like to move away from that. Blaming the dark (and foreign) forces of capitalism for the series of disasters associated with mass tourism is too easy and omits thinking about the specific role and agencies within the Mediterranean that helped shape tourist development. It is necessary to develop historical and sociological accounts that explore the actual workings of mass tourism so that we can get a sense of the diversity and multiplicity of this tourist phenomenon.

In recent years a number of monographs have appeared, although not always in English, that unpack the varying historical and geographical contexts for mass tourism. While often responding to local interests, these lines of research have the potential to shed light on the workings of mass tourism more generally. For instance the specifics of the Balearic Islands are explored in a series of works by Joan-Carles Cirer on Eivissa and Joan Amer on Mallorca, both in Catalan, which add to an existing literature that also includes the edited collection of La Caixa Foundation (2000) and the work of Rozenberg (1990). Instead of presenting tourism as an immutable, extra-historical development imported from the north, these accounts show mass tourism as a contested field, the shape of which is always provisional and constantly re-negotiated. While taking different perspectives, reading these monographs we get a sense that mass tourism is less homogeneous and straight forward than we thought. We also get a sense that mass tourism is not so much an example of cultural and social dependency as the Mediterranean route to Modernity. We hope that comparative histories as well as transnational accounts can emerge that look beyond case studies to follow the different links, routes and flows of mass tourism.

Doing the Med, Imagining the Med

If mass tourism produces times, spaces and bodies it also produces knowledge – though not of the kind generally recognized in academic accounts. Tourism is now one of the common frames for seeing and sensing the world with its own toolkit of technologies, techniques and aesthetic sensibilities for accessing the world and positioning ourselves in it (Franklin and Crang 2001: 8). Mass tourism is a major medium through which millions of Europeans have been able to know, experience and imagine the vast and fluid space of the Mediterranean. Chapters in this volume thus look at how mass tourism is framing Mediterranean space, by creating sacralized sites, by scripting places with different and often conflicting cultural narratives, re-ordering and reinterpreting a region divided on political and religious lines. Many of these framings draw upon discourses orientalism which, as Minca, Borghi and Haldrup show later in the book, still subtend tourist practices in and expectations of places such as Egypt and Morocco. The Mediterranean is often located and framed at the edge of Europe and if not part of the exotic orient at least half way to it. There is however no single tourist framing of the Mediterranean. Different *vacationscapes*, as Löfgren (1999) calls them, coexist, each of them with its own collective images, fantasies and everyday practices and routines, framing and scripting the Mediterranean differently, as a place to encounter civilization and as a place to get away from civilization. The differential scripts and imaginaries, freighted with historical legacies, that sustain new power relations and define place identities are not simply homogenized in mass tourism.

As well as a way of accessing the world, mass tourism is a way of producing meaning about the self. The majority of tourists go to the Mediterranean in search of a sunny place near the sea to relax and get away from it all. Not to find themselves in the popular sense, but certainly to produce a tourist subject that is transformed into a relaxed and liberated agent, and indeed to experience tourism as the demands to relax and enjoy. In focusing on the processes of becoming tourists, this book is moving away from individualist and instrumentalist conceptualizations of subjectivity. What is important for us it is not to understand the tourist subject but rather the always-relation practices of subjectification (Thrift 1997). Human agency is a relational effect generated by a decentred network of heterogeneous, interacting components. The chapters of this book look at the Mediterranean as a space for, and tourism as a technology of, the transient re-definition and multiplication of the self. The most extreme cases of transformation are linked to party destinations such as Eivissa and Falaraki, where people travel to take part in round the clock dance parties and loose themselves in the alcohol, the music and the drugs, Dan Knox in Chapter 8 develops insights into the cases of Faliraki in Greece and Ayia Napa in Cyprus. In these two cases the Mediterranean emerges as a site where tourists can explore the limits of hedonism and experiment with novel ways of transformation of the self and the body. Nevertheless cases like these are rather exceptional, the continuous de-composition and re-composition of the self that characterize mass tourism is more often modest and mundane. For many tourists, coastal holidays are a travel-inward to the world of the family as much as a voyage away. Being together and reinforcing family bonds are the basic ambitions of many Mediterranean holidays – an arena for 'doing family' that both demands and produces a variety of familial performances. In this case the Mediterranean emerges as a 'theatre of life' in which tourists can cultivate new aspects of their familial and personal identities. Performing a united family and becoming a child again are some of the pleasures and pressures of mass tourism. The significance of the family is evident in tourist photography, which tend to evoke a happy, stable and united family (Haldrup and Larsen 2003, Crang 1999).

In looking at the production of knowledge in mass tourism, it is not our intention to make yet another study of tourist representations, but to develop insights into the tourist process of signification. A sense of mass tourism as a practice is therefore central to this book. This collection develops approaches to mass tourism which are sensitive to social practices and embodied performances. We are interested in the doing of the Mediterranean as much as in its imagining. Mass tourism takes shape and gains expression in shared experiences, everyday routines, fleeting encounters, embodied movements, practical skills, collective dreams and sensuous dispositions. It is a social accomplishment, which has to be learned practised and developed and which enrols a variety of material elements in various places (from pre-holiday tanning parlours, to shopping trips to rituals of beach behaviour). In developing a sense of mass tourism as open practice we are not proposing the elimination of representations, meanings or any sort of symbolic activity from the study of mass tourism, but their integration within a single world, which is both open and pragmatic. There is a need to develop a broader scope of mass tourism that integrates practices and fantasies, objects and dreams, moving simultaneously in a material and utopian terrain. As Crang points out, 'it is not about what representations *show* so much as what they *do*' (2006: 48). Even in those environments surrounded by the fleeting, subjects are actively engaging with the world in a creative and productive manner. Reframing mass tourism as creative, embodied and performative, this collection takes on the challenge of reintroducing a sense of sensuality and enjoyment to the costal resort, thus re-enlivening a set of tourist geographies which are often left bare by the rush to produce theoretical order and economic diagnoses. This book explores a number of articulations of practice and subjectivity tied to mass tourism, including those related with photography, the coastal hotel and nightlife. It illustrates how practices of tourism bleed back into 'everyday life', and how time in the Mediterranean relates to technologies of the self, imaginations and cultural understandings.

The heterogeneous practices of tourism are made possible by complex materialities through many networks, connections and disconnections. Tourist practices and subjectivities are the product of complex interactions of human and non-human agencies. Tourists are integrated in assemblages, of things, technologies and places. There are multiple interdependencies between mass tourism, architecture and different technologies unfolded in our lives, including technologies of representations. Chapters will explore these complex interdependencies, thus demonstrating how tourists are constantly attending to the material world, looking using, buying and even making things with their hands.

Outline of the Book

This edited collection offers a series of insights into some of the key sites of mass Mediterranean tourism. It contains nine chapters covering Morocco, Egypt, Turkey, Cyprus, Greece and Spain as well as the intersection between northern Europe and the Mediterranean. The book is organized thematically as well as geographically. The early chapters focus on the southern shore, bringing in the colonial and exotic legacies underpinning Mediterranean tourism. The later chapters are concerned with the more hedonistic forms of mass tourism which are predominant in the northern shore. The intersection of the material with the representational and the performative is a major concern in all the contributions. However the early chapters place more emphasis on the complex visualizations and imaginary of the Mediterranean, whereas the later chapters switch the focus to social practices and embodied performances. All chapters report from recent and ongoing field research in the Mediterranean offering up to the minute material and are engaged with the critical interpretation of tourism. The book starts with this introductory chapter that makes a critical review of social and geographies research on mass Mediterranean tourism, identifying neglected areas of inquiry as well as an emerging research agenda.

The first two chapters of the book reflect on the more 'cultural' and 'exotic' dimension of Mediterranean tourism. In Chapter 2 Minca and Borghi look at the new forms of mass cultural tourism that are rapidly transforming Morocco into a key cultural destination in the Mediterranean. The cultural shift of Moroccan tourism is based on the re-staging of the colonial for the masses and the promotion of Morocco as a secure and easy-to-reach exoticism. The process is part of a strategic vision of rapid growth that proposes the rewriting of the imaginary of Morocco through an orientalist eye, refashioning the tourism geographies inaugurated by the colonial French. Minca and Borghi focus on the tourist colonization of Marrakech's Medina. They unpack the translation of its orientalist image into a new landscape aesthetics and a series of mundane practices. By highlighting the distance between the exotic aura of tourism promotion and the banality of many tourism practices, the chapter deflects usual criticism aimed at mass cultural tourism for its reliance on false and static tropes. When translating orientalist discourses into mundane practices, stereotyped representations and performances of colonial aesthetics are challenged and even disrupted.

The ambivalence between cosmopolitanism and orientalism in contemporary tourism is the main theme of Michael Haldrup's Chapter 3. The chapter examines the significance of tourism as a mundane form of experience of the global and the other, drawing on ethnographic research on Danish tourists visiting Egypt. Haldrup emphasizes the importance of mundane banalities for the study of tourist performances. Bringing together the exotic and the familiar, the chapter explores the multiple ways tourist performances feed into everyday life of people. Instead of 'banalizing' tourism, Haldrup calls attention to the banal ways tourist practices and experiences are implicated in the emergence of contemporary cultures of banal cosmopolitanism and orientalism. The chapter also emphasizes the need to trace out the routes that connect the context of home and away. Tourism is a culture of circulation that cannot be easily contained and localized. By discriminating between three different modes of banal cosmopolitanism Haldrup points towards the significance of less reflexive and more embodied responses to the penetration of everyday life by global mobilities.

The following chapter, which takes us to the Greek Island of Kefalonia, the setting of the book and film of *Captain Corelli's Mandolin*, develops a case study of middlebrow popular tourism in the Mediterranean. Crang and Travlou's Chapter 4 unpacks the production and enactment of tourist imaginary in the island, drawing attention the intersection between tourism, film and literature. The chapter explores how the film and the book have refigured the tourist image of the island by replaying the myth of the Edenic Beach as well as ongoing discourses of Greekness and Mediterraneanness. The movie and book of *Captain Corelli* wrap the island in a romantic imaginary, which is part of the currency through which the tourist appeal of the island is constructed over and against the implied carnivalesque hedonism of other destinations. The chapter also explores how this image is played out in practices of tourist consumption on different beaches. In making such connections between tourist images and practices, the absences are as revealing as the presences. *Captain Corelli* can hardly be missed and yet it is also curiously hard to actually find it. *Captain Corelli* is a phantasm in the island. Its substantiality is diffuse and yet undeniably present.

Chapter 5 is the first to reflect on 'sun, sea and sand' forms of mass tourism. In this chapter Pau Obrador examines the coastal hotel, one of the most basic institutions of mass tourism in the Mediterranean. The chapter reflects on the nature of social relations in the highly commodified and fleeting environments of mass tourism drawing attention to the everyday relations and practices through which this space is inhabited by tourists. Obrador considers the coastal hotel in terms of hospitality and affect, opening up mass tourism to new ways of thinking the social that are less hostile to this tourist phenomenon. An ethnographic account of two hotel pools in the island of Menorca explores the various elements that make up the social fabric of the pool including conviviality, domesticity, hostility and the gaze. The discussion brings to the fore the complexity and liveliness of the social life of the pool. By the pool light touch forms of sociality coexist with more enduring social forms that combine an ideological and affectual dimension.

The cultural geographies of 'sun sea and sand' in the Mediterranean are highly diverse incorporating a myriad of configurations, some of which remain largely unacknowledged. In Chapter 6, Javier Caletrío examines one of the alternative geographies of mass tourism and the beach that coexist along with the more visible landscapes of the package tour. This alternative tourist geography of the Mediterranean is a largely domestic phenomenon that centres on the figure of *veraneante* translating as 'those who regularly spend the holiday there'. Focusing on the Costa Blanca in Spain, the chapter highlights the significance of the familiar in the creation of the Mediterranean landscape. It is a form of vacationing that centres on the re-encounter of place and people and where meeting significant others is critical. It is an enduring tourism geography punctuated by quotidian rituals, ways of encountering that have to be learned. The chapter also emphasizes the embodied and material ways in which geographies and landscapes are mobilized. For the *veraneante* the sense of familiarity and intimacy stems from a practical engagement with the sensualities of the beach and the seaside in company of others. The beach of the *veraneante* is a landscape of memory, heavy with time that speaks of an investment in a sense of place.

The extraordinary development of mass tourism in places like the Costa del Sol has led to new touristic mobilities with international visitors also becoming more sedentary. Residential tourism, which Karen O'Reilly examines in Chapter 7, is one of the most significant of them. A series of contradictions characterize these transnational communities settling in the Costa del Sol. Residential tourism entails a subtle and continuous balancing act between residence and tourism, home and away, utopia and reality, which has to be managed, practiced and performed on a daily basis, specially through the act of being both host and guest. O'Reilly regards these contradictions as essential to the lives of residential tourists enabling them to live their lives as a permanent escape. Their lifestyle works as long as participants remain in but not in, home but not home neither here nor there. O'Reilly incorporates materiality in her cultural account of the phenomenon examining the material cultures that are woven into the everyday life of British residents in Spain and their co-creation through transformation and performance.

Chapter 8 develops insights into the more hedonistic side of mass tourism. Drawing on participant observation in the resorts of Faliraki in Greece and Ayia Napa in Cyprus, Dan Knox examines the often vilified mass youth tourism, making a plea for deeper and more nuanced understanding of the phenomenon. He looks at both the signifying practices of those participating in clubbing holiday and the representations of such holidays and how the circulation of those representations fuels the continued consumption of youth tourism products. The chapter is primarily concerned with the relationship between mass tourism and the everyday. Knox argues that youth tourism is not so much a break from general life experiences as a period of increased activity, a spectacular manifestation of the everyday. Representations consumed in advance such as media performances and club soundtracks heighten a sense of expectation while providing the necessary knowledge to become insiders of the scene. An inflection of Bourdieu's account of cultural capital to become subcultural capital is central to the chapter. Like more sophisticated forms of tourism, youth tourism is also a way of gathering valuable experience that has value within particular social field.

The book finishes with a commentary on the Mediterranean on the age of mass tourism, following the spirit of French historian Fernand Braudel. This commentary emphasizes the significance of mass tourism in shaping and reshaping the region. Mass tourism has become the very fabric of the Mediterranean, conferring to the region a new economic and social centrality. It is the Mediterranean route to post-modernity.

References

Abram, S., Waldren, J. and Macleod, D.V.L. 1997. *Tourists and Tourism*, London: Berg Publishers.

Aguiló, E., Alegre, J. and Sard, M. 2005. The persistence of the sun and sand tourism model. *Tourism Management*, 26(2), 219–31.

Amer, J. 2006. *Turisme i Política: L'Empresariat Hoteler de Mallorca*. Palma: Documenta Balear.

Apostolopoulos, Y., Loukissas, P.J. and Leontidou, L. 2001. *Mediterranean Tourism: Facets of Socioeconomic Development and Cultural Change*. London: Routledge.

Augé, M. 1995. *Non-Places: Introduction to an Anthropology of Supermodernity*. London: Verso.

Bærenholdt, J.O., Haldrup, M., Larsen, J. and Urry, J. 2004. *Performing Tourist Places*. Aldershot: Ashgate.

Bakhtin, M., 1984. *Rabelais and His World*. Indianapolis: Indiana University Press.

Bauman, Z. 2000. *Liquid Modernity*. Cambridge: Polity Press.

Bennett, T. 1983. A thousand and one troubles: Blackpool Pleasure Beach, in *Formations of Pleasure*, edited by F. Jameson et al. London: Routledge, 138–55.

Billig, P.M. 1995. *Banal Nationalism*. London: Sage.

Boissevain, J. 1996. *Coping With Tourists: European Reactions to Mass Tourism*. Oxford: Berghahn Books.

Borch, C. 2006. The exclusion of the crowd: the destiny of a sociological figure of the irrational. *European Journal of Social Theory*, 9(1), 83–102.

Bramwell, B. 2003. *Coastal Mass Tourism: Diversification and Sustainable Development in Southern Europe*. Clevedon: Channel View Publications.

Carter, S. 2007. *Rise and Shine: Sunlight, Technology and Health*. Oxford: Berg Publishers.

de Certeau, M. 1984. *The Practice of Everyday Life*. Berkeley, CA: California University Press.

Cirer, J. 2004. *De la Fonda a l'Hotel*. Palma: Documenta Balear.

Corbin, A. 1994. *The Lure of the Sea: The Discovery of the Seaside in the Western World, 1750–1840*. Berkeley, CA: University of California Press.

Crang, M. 1999. Knowing, tourism and practices of vision, in *Leisure/Tourism Geographies: Practices and Geographical Knowledge*, edited by D. Crouch. London: Routledge, 136–53.

Crang, M. 2006. Circulation and emplacement: the hollowed-out performance of Tourism, in *Travels in Paradox*, edited by C. Minca and T. Oakes. Lanham, MD: Rowman and Littlefield, 47–64.

Crang, M. 2009. Moving places, becoming tourist, becoming ethnographer, in *Mobilities: Practices, Spaces, Subjects*, edited by T. Cresswell and P. Merriman. Farnham: Ashgate.

Crick, M. 1992. Ali & Me: an essay in street corner anthropology, in *Anthropology and Autobiography*, edited by J. Okely and H. Callaway. London: Routledge, 175–92.

Crouch, D. (ed.) 1999. *Leisure/Tourism Geographies: Practices and Geographical Knowledge*. London: Routledge.

Debord, G. 1973. *Society of the Spectacle*. Detroit: Black & Red Books.

Desmond, J. 1999. *Staging Tourism: Bodies on Display from Waikiki to Sea World*. Chicago, Il: University of Chicago Press.

Edensor, T. 2001. Performing tourism, staging tourism: (re)producing tourist space and practice. *Tourist Studies*, 1(1), 59–81.

Franklin, A. 2003. *Tourism: An Introduction*. London: Sage.

Franklin, A. and Crang, M. 2001. The trouble with tourism and travel theory. *Tourist Studies*, 1(1), 5–22.

Foucault, M. 1978. *The History of Sexuality, Vol. 1: An Introduction*. New York: Vintage Books.

Galani-Moutafi, V. 2000. The self and the other: traveller, ethnographer, tourist. *Annals of Tourism Research*, 27(1), 203–24.

Guattari, F. 1992. *Chaosmosis: A New Ethico-Aesthetic Paradigm*. Indianapolis: Indiana University Press.

Haldrup, M. and Larsen, J. 2003. The family gaze. *Tourist Studies*, 3(1), 23–46.

Huizinga, J. 1998. *Homo Ludens: A Study of the Play-Element in Culture*. London: Routledge.

Inglis, F. 2000. *The Delicious History of the Holiday*. London: Routledge.

Koshar, R. 1998. 'What ought to be seen': tourists' guidebooks and national identities in modern Germany and Europe. *Journal of Contemporary History*, 33(3), 323–40.

Le Bon, G. 2002 [1895]. *The Crowd: A Study of the Popular Mind*. New York, Dover Publications.

Lenček, L. and G. Bosker 1998. *The Beach: The History of Paradise on Earth*. London: Secker & Warburg.

Löfgren, O. 1985. Wish you were here! Holiday images and picture postcards. *Ethnologia Scandinavica*, 15, 90–107.

Löfgren, O. 1999. *On Holiday: A History of Vacationing*. Berkeley, CA, University of California Press.

Fundació La Caixa. 2000. *Welcome! Un segle de Turisme a les Illes Balears*. Barcelona: Fundació La Caixa.

Massey, D.B. 2005. *For Space*. London: Sage.

Minca, C. and Oakes, T. (eds) 2006. *Travels in Paradox: Remapping Tourism*. Lanham: Rowman & Littlefield Publishers.

Montanari, A. and Williams, A.M. 1995. *European Tourism: Regions, Spaces, and Restructuring*. Oxford: Wiley

Morris, M. 1988. Banality in cultural studies. *Discourse*, 10(2), 3–29.

MVRDV 2000. *Costa Iberica: Upbeat to the Leisure City*. Barcelona: Actar.

O'Reilly, K. 2000. *The British on the Costa Del Sol: Transnational Identities and Local Communities*. London: Routledge.

Pack, S.D. 2006. *Tourism and Dictatorship: Europe's Peaceful Invasion of Franco's Spain*. New York: Palgrave Macmillan.

Risse, M. 1998. White knee socks versus photojournalist vests: distinguishing between travellers and tourists, in *Travel Culture: Essays on What Makes Us Go*, edited by C.T. Williams. Westport, CT: Praeger, 40–50.

Rojek, C. 1993. *Ways of Escape: Modern Transformations in Leisure and Travel*. London: Palgrave Macmillan.

Rozenberg, P.D. 1990. *Ibiza, una Isla Para Otra Vida*, Madrid: Centro de Investigaciones Sociológicas.

Seigworth, G. 2000. Banality for cultural studies. *Cultural Studies*, 14(2), 227–68.

Selwyn, T. 1996. *The Tourist Image: Myths and Myth Making in Modern Tourism*. Chichester: Wiley.

Sheller, M. 2004. Demobilizing and remobilizing Caribbean paradise, in *Tourism Mobilities: Places to Play, Places in Play*, edited by M. Sheller and J. Urry. London: Routledge, 13–21.

Sheller, M. and Urry, J. 2004. *Tourism Mobilities: Places to Play, Places in Play*. London: Routledge

Thrift, N, 1997. The still point: resistance, expressive embodiment and dance, in *Geographies of Resistance*, edited by S. Pile and M. Keith. London: Routledge, 124–51.

Turner, L. and Ash, J. 1975. *The Golden Hordes: International Tourism and the Pleasure Periphery*. London: Constable.

Urbain, J.D. 2003. *At the Beach*. Minneapolis: University of Minnesota Press.

Urry, J. 1990. *The Tourist Gaze: Leisure and Travel in Contemporary Societies*. London: Sage.

Vasudevan, A. 2006. Experimental urbanisms: Psychotechnik in Weimar Berlin. *Environment and Planning D: Society and Space*, 24(6), 799–826.

Vilarós, T.M. 2005. Banalidad y biopolítica: la transición y el nuevo orden del mundo, in *Desacuerdos*, edited by MACBA. Barcelona: MACBA, 25–56.

Waldren, J. 1996. *Insiders and Outsiders*, London: Berghahn Books.

Walton, J.K. 2000. *The British Seaside: Holidays and Resorts in the Twentieth Century*. Manchester: Manchester University Press.

Chapter 2

Morocco: Restaging Colonialism for the Masses

Claudio Minca and Rachele Borghi

Introduction

Jamaa el Fna square, Marrakech, on a hot May afternoon: two water sellers (*garraba*) in traditional attire (Figure 2.1) pose for the tourists' cameras, resplendent in their colourful clothes, their weathered leather sacks full of water. This performance often includes a photo with the tourist pretending to drink the water taken from the sacks. With the *mise en scène* over, the tourist usually drops a coin into their purse and, sometimes, begins chatting with the water-seller that just moments previously was the centre of the performance (Figure 2.2). This passage between the 'front-stage' and the 'back-stage', where the informal chat takes place, appears effortless, as though it too were part of what Goffman (1959) would term 'the representation'.

The water sellers have for long been important figures in the global iconography that inscribes Morocco and the culture(s) of its mass tourism; an iconography that, by now, is part of the European collective travel imaginary. The water sellers also often appear in the promotional materials produced by the Moroccan tourist authorities. Their performance represents a sort of tourist 'tradition', re-enacted daily in the Jamaa el Fna. Yet what is most interesting in this scene is the fundamentally banal nature of the water sellers' tourist experience; a banality not only confirmed by the commercial transaction it entails, but also by the fact that the water sellers, after the 'staging' process, are often treated by the tourists as people like any other, thus deprived of any 'Oriental' aura or exoticism.

In this chapter, we intend to highlight precisely this (only apparent) distance between the 'exotic' aura that tourist promotion assigns to the 'cultural experience of Morocco' and the banality of many tourist practices. We focus our comments on Marrakech in particular, often presented as the iconic site of Moroccaness. We suggest that such experiences and practices are, in fact, also the result of the repetitive nature of many tourist performances: repetition in the hosts's performances – who meet different tourists but follow the same rituals/codes many times every day; and repetition in the tourists's own practices – who collect very similar experiences and play similar roles in many different moments and places while travelling (see Rojek 1996, Coleman and Crang 2002, Oakes 2005, 2006). More broadly, we focus on the (banality) of the Orientalist images adopted

**Figure 2.1   Water seller (garraba) in Jamaa el Fna, Marrakech
               (photo by Rachele Borghi)**

**Figure 2.2   Water seller and a tourist chatting after the ritual
               photo-performance (photo by Claudio Minca)**

to promote Morocco, on the one hand, and the actual translation of those latter into 'real' geographies of tourism, into a newly spatialized colonial nostalgia that is literally transforming the face of some *medinas* and other tourist spaces.

The recent evolution of Moroccan mass tourism towards a 'cultural turn' of sorts (see Borghi 2004, Filali 2008) has, in fact, been matched by growing attention on the part of the national government to the economic and political role that tourism can potentially play in the future development of the country. The Tourist Master Plan launched by the Moroccan Government during the *Assise Nationale du Tourisme* held in Marrakech in January 2001 aims for rapid growth in the number of visitors and the consolidation of Morocco as key 'cultural tourism' destination in the Mediterranean. The Master Plan reflects a revolutionary vision for Moroccan tourism that by 2010 should reach the magic threshold of 10 million visitors a year, thanks to a strategic diversification of its attractions and itineraries, the improvement of access and transportation, and the realization of new hotels and resorts (Daoud 2001: 28). While also making reference to existing competition with other Mediterranean destinations, such as Turkey, Tunisia and Egypt, the Plan presents tourism as a key factor in the enhancement of the prestige and the visibility of Morocco on the international scene more broadly (Franco 1996: 26, Filali 2000: 10, El Amrani 2001: 26). The aim of this 'new vision for tourism' is, indeed, creating a new 'culture of hospitality' able to accommodate mass arrivals from Europe, new, more sophisticated, expressions of colonial aesthetics, and new forms of secure (geopolitically speaking) and easy-to-reach Oriental exoticism.

The new Plan has proved relatively successful, doubling the number of visitors in just a few years (reaching about 4.1 millions in 2006).[1] Combining an aggressive campaign to encourage further international investment in Marrakech (designated as Morocco's 'showcase' on the international scene) with the signing of an 'Open Skies' agreement with European carriers, the Plan also encouraged the promotion of a new international image based on the restaging of the country's colonial past for the new mass tourists. The campaign found fertile ground within the European market, long prepared to experience Morocco as the ultimate destination for the materialization of a sophisticated colonial nostalgia.[2]

In this chapter, we will reflect on the selling of such 'colonial nostalgia' and on some of its practical implications for contemporary Moroccan mass tourism. What we will try to highlight is how, in some places and some moments in time, tourist experiences are translated into the production of a set of (often banal) practices that often (playfully) challenge the global(ized) representations that inscribe and promote those very practices (see Bærenholdt et al. 2004). By drawing attention to such instances, we hope to, on the one hand, deflect (at

---

1   See   http://www.tourisme.gov.ma/francais/5-Tourisme-chiffres/ArriveeTouristes. htm [accessed: 19 June 2008].

2   A trend that has also been observed in other former colonies, see Bissell 2005; Gregory 2001; Mahr 2007; Peleggi 2005; Rosaldo 1989.

least in part) the usual criticism levelled at mass cultural tourism – that is, its presumed reliance on 'false' and static (neo)colonial images of other cultures (see Minca 2007); on the other, we wish to underscore how such instances often support a much more mundane interpretation of tourism and of the culture that they help produce (see Crang 1999, 2004, 2006; Crouch 2004, 2005; Edensor 2001, 2006). The tourist 'staging of the colonial', in all its banality and sometime vulgarity, contributes in fact to mobilizing people, capital and images, while also materially transforming places and the meanings attached to them (see Henderson and Weisgrau 2007).

Following a brief genealogy of mass tourism in Morocco, we focus our attention on the new 'cultural' turn in Moroccan tourism. In particular, we highlight some of the processes of the restaging of the colonial for the masses in the Marrakech medina, building on our previous research on this topic (Borghi 2008a, Minca 2006, Borghi and Minca 2003). Tourism in Marrakech and, more specifically, the tourist colonization of the medina (see, for example, Escher et al. 2001a, 2001b) provides a particularly important focus for our central argument, also because the implications that these processes have and are likely to have in the future development of mass tourism in the rest of the country.

The opening up of Moroccan skies to low-cost European airlines a few years ago has, de facto, brought Marrakech much closer to European tourist markets, favouring the growth of new forms of cultural mass tourism. In Marrakech, and especially in its increasingly gentrified medina, the objectification and sacralization of certain aspects of Moroccan life assumes almost grotesque manifestations, reflected in the tourists's own performances. This is due both to a grossly oversimplified understanding of 'Arab life' and to the fact that tourists must depend on local 'mediators' and local organizations to help them negotiate complex urban spaces that would otherwise be extremely difficult to approach or even reach. As this chapter will suggest, in such spaces aestheticized colonial nostalgia is coupled with an extraordinary degree of banality in tourist practices. The way in which such massified banal practices dialogue with representations produced and promoted by the media and by tour operators is an intriguing and partially overlooked field of investigation, at least in the Moroccan context.

L'Appel du Maroc

In the early nineties, the French advertisement company *Publicis* was commissioned by the Moroccan Government to create a new promotional campaign entitled '*Maroc: l'éblouissement des sens*', a campaign that was strongly marked by appeals to exotic and Orientalist understandings of the 'Morocco experience' (Ejeux 1993). Such understandings are recalled today by almost all Moroccan tour operators and dedicated websites, as well as publications produced by the Ministry of Tourism and the *Office National Marocain du Tourisme*. Morocco is presented as an exotic yet at the same time easy-to-reach destination, characterized by an 'uncontaminated nature', picturesque landscapes, labyrinthine medina(s), and a unique 'culture of hospitality':

> As Western Europe's closest African and Arab neighbor, Morocco enjoys regular and very convenient air links across the World. It also boasts a rapidly expanding modern telecommunications network, hotels of international standard and the kind of infrastructure that makes Morocco one of the most highly developed and politically stable countries in Africa.

> At the same time, Morocco is a country steeped in history and traditional culture that has lost none of its authenticity despite its close links with the West.

> Blessed with some of the most stunning landscapes of any country in the world, Morocco has justly become a firm favorite amongst film makers, photographers, travel planners and, most importantly, conference organizers.[3]

Alizes Travel, a tour operator based in Casablanca claims to offer, instead, a 'Morocco where dreams come true'. Its English-language catalogue contains a series of washed-out, hazy images, presumably evoking the country's 'dream-like' atmosphere. The pictures that embellish the catalogue are largely of Marrakech's 'magical' landscapes (the city's ochre-red walls, Jamaa el Fna square, the snowcapped Atlas Mountains and so on). In the background lie the contours of the country, marked with the key destinations of their 'tours' (Tangiers, Rabat, Casablanca, Marrakech itself, Agadir, Laayoune and so on).

It is interesting that here, as in most of the brochures we examined, the images convey a rather ambivalent message, for references to 'upscale' modern facilities are frequently accompanied by the evocation of the work of Orientalist painters like Delacroix, Majorelle and Matisse and, in general, an Orientalist, colonial imaginary. The dominant trope is, in fact, the wholly-colonial '*L'Appel du Maroc*'[4] – a trope that draws on a long tradition of itinerant cultural tourism, inaugurated already by the French during the days of the Protectorate, now revived and reinterpreted within the contemporary Moroccan restaging of the colonial for the masses.

It is towards the end of the 1800s that the 'Orientalist gaze' shifted its focus from cities like Cairo and Shanghai, turning its eyes towards Morocco (Rivet 1984: 97), that soon comes to be perceived as the most fascinating materialization of the Oriental ideal. This 'vision' of Morocco was actively promoted by painters and writers, but also by the *savants* of the time (Borghi 2008a, Rivet 1984). These latter, though primarily engaged in the 'translation' of Morocco into the language

---

3   Travel Link, a Marrakech-based tour operator, www.travellink.ma [accessed: 19 June 2008].

4   This trope was, indeed, the focus of a recent volume edited by Rondeau (2000), published by the Paris-based *Institute du Monde Arabe*.

of scientific discovery and colonial enterprise, were at once concerned for the potential loss of its 'magic aura' that modernization might bring:

> Our love for the Morocco of old, our joy in travelling through its landscapes, in experiencing the fascination of its Islamic civilization, primitive yet nonetheless captivating, all these are marked by an underlying melancholy, an underlying realization: so many of these enticements, so many of these emotions will soon disappear (Ladreit de Lacharrière 1911, cited in Rivet 1988: 20)

Such a vision of Morocco was also sustained by a new '*marocanisant*' literature, inspired in large part by the scenes portrayed by Delacroix (1893, 1999; see also Turco 1995). This literature emerges at the end of the nineteenth century and soon becomes very popular among the French and European readership more broadly, a readership irresistibly drawn to exotic, 'Oriental' experiences (see Potier 2006). Morocco appears to represent, for many, the last bastion of Oriental beauty and Islamic civilization, able to offer to the European traveller the profound sense of *dépaysement* they yearned for (Boëhm 1993). The first travel guide to Morocco was written in 1889 by M. de Kerce, at the time the editor of a Tangiers-based magazine bearing the evocative title *Le réveil du Maroc* (Lebel 1936: 347). The nineteenth century pioneers of modern tourism were mainly artists, explorers and missionaries, travelling individually. Nonetheless, their accounts of Morocco (see for example de Foucauld 1888, Loti 1889, De Amicis 1913) became of crucial importance in drawing a first (European) map of the potential tourist 'resources' of the country, a map that the French Protectorate (established in 1912) would adopt as a key referent for its own (tourist) re-interpretation of Moroccan places and landscapes (see Wharton 2003).

Despite the relative political instability and lack of military 'pacification' of many parts of the Protectorate in its early days, the French Governor, *Resident General* Marechal Herbert Lyautey decided to launch a campaign for the development of international tourism in the country on a relatively large scale. The role of tourism, in Lyautey's vision of Morocco,[5] was particularly important for two reasons: firstly, by attracting European (mainly French) visitors, tourism would help incorporate the 'new' Morocco into European geographies of culture and nature; secondly, this form of 'appropriation' in many ways reflected Lyautey's own ideology of the 'valorization' of the *patrimoine* of the colonized country, confirming the (political) legitimacy of his ambitious colonial vision for Morocco (see Morton 2000, Rabinow 1989). The arrival of (an albeit limited number of) tourists in the early decades of the 1900s was thus seen as an important mark of the successful pacification of this at times 'rebellious land', while also proving justification for further (military) 'stabilization', as well as the establishment

---

5   On this vision see Barthou 1930; Benoist-Mechin 2007; Doury 2002; Hoisington 1995; Laprade 1934; Lyautey 1995; Maurois 1931; Rabinow 1989; Rivet 1980, 1988, 2002; Venier 1997.

of better communication networks   and pre-disposition of some highly representative sites in Morocco's key cities for the nascent tourist market.

The first forms of modern tourism in Morocco developed, indeed, in the 1920s (see Cattedra 1990) and it was in these early stages that a specific geography of tourism, closely shaped by the colonial desire for (and military control of) Morocco, began to take shape; a geography that would remain essentially unchanged all through the twentieth century. The first all-inclusive tour was organized in 1920: starting in Bordeaux, over the space of three weeks it took tourists through Casablanca, El Jadida, Marrakech, Meknes, Fes, Oujda and Oran, ending in Algiers. The first tourist resorts, on the other hand, came to be realized within a securitized territorial triangle along the Atlantic coastline, marked by the pivotal roles of Tangiers, Fès and Marrakech, that soon became the most popular Moroccan destination among European travellers (Weisweiller 1932: 104).

Lyautey, in fact, was careful to lay the ground for Marrakech's emergence as a successful tourist destination: his comprehensive urban plan created majestic boulevards, beautiful garden spaces and prestigious colonial buildings, built in a Moresque style. Indeed, the Resident General's architects used a sophisticated play of perspectives in creating Marrakech's *Ville Nouvelle* in order to emphasize views of the famous Koutobia minaret, the city's red walls and the Atlas Mountains (see Borghi 2008b, Clément 1994, Minca 2006). Travellers to Marrakech (and Morocco in general) during the first decades of the Protectorate also lamented, however, the lack of adequate services and hotels able to attract and accommodate a rich and sophisticated clientele. As one commentator noted,

> The hotels [in Marrakech] lack the comforts expected by a rich tourist clientele, used to the palaces of Italy or Egypt [...]. We can only hope that one day Marrakech, too, will boast one of these hotels, a veritable colonial palace, set in a large garden full of exotic trees, palms and flowers, with apartments specially adapted for life in a climate that is mild in the winter and very hot in the summer months.

> At that point, Marrakech will become truly the tourist destination I dream of. The railways will also join Guéliz,[6] carrying the suitcases, trunks and hat-boxes of elegant women who will play tennis on courts shaded by palms in the December sun, oblivious to all the work that has gone into conquering, conserving and constructing for them this winter paradise (Dugard 1918: 181–2).

The dream soon came true. In 1921, Lyautey inaugurated the majestic Mamounia Hotel that rapidly gained an extraordinary international reputation (Compagnie

---

6   Marrakech's *Ville Nouvelle*, created by Lyautey's architects (see Borghi 2008a; 2008b; Clément 1994).

Figure 2.3   La Mamounia (photo by Rachele Borghi)

des chemins de fer au Maroc 1921: 1).[7] With the opening of the Mamounia (see Figure 2.3), Marrakech became a prestigious destination and a gathering point for the international 'jet-set' of the day, hosting many important events (Bauchet 1954: 44). That very same year, the French publisher Hachette released the third edition of the *Guide Bleu* to Morocco (the first one having been published in 1918); the initiative was highly praised by the Protectorate authorities and, indeed, the preface was written by Marechal Lyautey (1918: 1) himself, noting that the Guide 'provided an exceptional opportunity to publicize a country destined to be an important tourist destination'.

In 1937, after almost two decades of growth of the tourist sector, the *Office Chérifien du Tourisme* was created. Its purpose was 'to study all tourism related questions, both with and outside of Morocco, and seek the best possible means to promote its development' (Hillali 2007). The Office was shut down with the outbreak of the Second World War, but reopened in 1946 with a new name – the *Office National Marocain du Tourisme* – an institution that survived the transition

---

7   The Mamounia remained, for many years, the winter holiday spot of choice of the international 'jet-set'. It is said to have been a favourite of Winston Churchill (see Mourad 1994).

towards independence and that is still active today.[8] The legacy of the colonial period was not limited, however, to such institutions or even the transformation of tourism into an important economic resource for the country. Lyautey's vision – and projects – comprehensively re-wrote Morocco for the European travelling eye, re-inscribing the country as an exotic destination: arguably still a decisive element in attracting mass tourism to Morocco today. With the end of the Protectorate in 1956, Morocco inherited both the facilities and the 'cultures' of tourism created by the French colonial project over the space of four decades.[9] These tourist cultures (and infrastructures) strongly influenced the policies of the first decades of the independent Morocco state – and are partially still with us today (Filali 2008).

'Mature' Tourism Geographies

As it enters the twenty-first century, Moroccan tourism is characterized by a series of 'mature' tourism geographies consolidated in the 1980s and 1990s, both in terms of the management of its attractions and hospitality industry, as well as a distinct collective imaginary that continues to attract European visitors. These geographies have played and continue to play today an extremely important role for the economic and political development of the country as a whole, and reverberate in intriguing ways in the 'products' that Morocco 'crafts' for its tourist masses (see Berriane 2003).

The Moroccan Government launched its first initiatives in the tourist sector just a few years following independence. It was the Mediterranean coast that received most of the early investment, a decision that had important implications for domestic tourism, but that never seriously impacted international arrivals (Hillali 2005). Morocco was – and remains today – an 'exotic' Arab and African destination for the European market, and its limited Mediterranean tourist development could never actually compete with other countries better equipped for mass beach tourism (Lozato-Giotart 2008). Nonetheless, by clearly focusing on tourism as a strategy of economic development already in those early days, the Moroccan Government made a decisive political choice; that is, to tie the country's economy to the 'free market, liberal bloc', at a moment in time when many other African countries were instead opting for variants of state-socialism. In this respect, it is particularly significant that in 1964, a group of experts from the World Bank was invited to visit the country and to draw up a plan for its future economic development. The Bank's experts agreed that Morocco should abandon large industrial projects in favour of developing its tourism sector, seen as capable of attracting more foreign investment and revenues, while creating jobs through

---

8   See http://www.tourisme.gov.ma.

9   In 1953, there were 265 hotels in Morocco, with approximately 7,500 rooms accommodating 253,000 tourists (Bélanger and Sarrasin 1996).

the 'valorization of the cultural and natural resources of the country' (Hillali 2007: 60–1).

From 1965 onwards, the combined effect of the encouragement of international organizations, the enthusiasm of multinational hotel groups and expectations of future investments on the part of the private sector, resulted in the launch of a series of plans for large scale tourism development. The 1965–72 Plan, in particular, inaugurated a new stage by explicitly identifying tourism as the second most important economic sector, after agriculture (ibid.). The Moroccan Government, having abandoned any grand ambitions of realizing a modern industrial apparatus, began to take an even stronger role in the management and promotion of the tourist sector. Tourism was seen, in fact, as a realm of development that required strong public intervention. Trying to learn from the Spanish experience of the 1960s, the Government's tourism development policies aimed at 'privileging mass tourism without depending too much on it, at diversifying its clientele, at regionalizing its effects and benefits and at improving its impact for the whole country' (Secréterait ... 2005, see also Galissot 1999).

Significant investment in the tourism sector was also seen as a form of compensation for some of the more peripheral areas of the country that were left out of the major economic developments following independence (mainly centered on the Tangiers-Casablanca-Rabat axis) (see Lozato-Giotart 1991). Such heavy-handed public intervention – ironically, quite reminiscent of many colonial strategies of economic management under the Protectorate – persisted up to 1978. The oil shocks of the early 1970s, the State's annexation of Western Sahara in 1975, and finally the collapse of phosphate prices in 1976–78, brought a significant downturn in the country's economic fortunes, resulting in the Government's partial withdrawal from the tourism sector and, in the decade following 1983, the implementation of a series of policies of 'structural adjustment' prescribed by the International Monetary Fund (Hillali 2007: 61).

Overall, while tourism confirmed some of its potentialities in the decades between 1960–1980, it never took off as a thriving sector able to support, with its direct and indirect impacts, the growth of the country's economy as a whole. This is reflected in some statistics from the period: starting from the very modest figure of 146,000 arrivals in 1960, the trend remained positive for more than a decade, reaching a million visitors in 1973. A long period of relative stagnation subsequently followed, with 1.3 million arrivals in 1984. Rapid growth resumed in the second half of the 1980s, culminating in the 3.2 million arrivals record figure in 1992. These numbers were to prove quite ephemeral, however, for the number of arrivals fell dramatically in 1995 as a consequence of the deadly terrorist attack in a Marrakech hotel in 1994 – recovering only partially by the turn of the century (2.4 million in 2000) (Département du Tourisme 2000).

Morocco's tourism geographies remained, by and large, those inherited from the Protectorate, with a leading role played by the traditional 'triangle' of the 'Imperial Cities' and with only limited impact of its Mediterranean coastal development. The launch of Agadir as the main – if not the sole – 'international beach destination' was partially successful, although based on a relatively standardized and lower-end tourist 'product', that soon began to suffer the competition of other 'new' Mediterranean destinations. Even the Moroccan desert South, given a further boost by the closing down of Algeria to international tourism in the 1990s, never achieved the figures and the impact predicted. By and large, the Moroccan authorities came to the realization that with the exception of the package tours headed for Agadir,[10] the country's real attraction was still very much its colonial legacy, with its corollary of exotic and Orientalist images that had not yet been 'exploited' in all their potential.

The affirmation, in the past few years, of a 'cultural' (and, implicitly, colonial) vocation of Moroccan mass tourism, together with a more general 'cultural shift' in Mediterranean tourism,[11] thus marked an important turning point in the planning strategies of Moroccan tourism. This shift has been crucial in the rapid increase in visitor s over the past few years, with a doubling of international arrivals (from 2.4 million in 2000 to more than four in 2006), but also a renewed valorization of what has by now become a genuine Moroccan 'tourist tradition' in the European collective geographical imagination(s).

The contemporary re-staging of the colonial for the masses thus responds to a very specific strategy based on a new landscaped aesthetics and a new set of regional and urban hierarchies that are often only implicit in the promotional rhetoric adopted to engage with tourists. These geographies however are not only well known by the tourists (who sometimes seem to follow, in their itineraries, a sort of invisible tread) but represent, moreover, a real mapping of the 'places' where (global) images of Moroccanness intersect with a set of concrete practices, of mundane and banal experiences, giving life, in this way, to new forms of cultural production, new performances, new spatialities. It is for this very reason that it is worthwhile reflecting on how the Moroccan authorities are attempting today to re-fashion these well established geographies, both in terms of images and in terms of the aesthetics and the functional organization of tourist spaces and places.

Re-staging Morocco (For the Masses)

10 January 2001 is often presented as the key turning point for Moroccan tourism and its new 'vision'. As previously noted, mass tourism had not taken off as hoped for in the post-independence decades, due to a number of problems. The 'Vision 2010' Master Plan hoped to address many of these: the lack of adequate air connections, weak transport infrastructures and facilities; a relatively poor quality and deregulated hotel industry; the harassment of tourists on the part of local *faux guides* in some mass tourism destinations; a strong dependence on the

---

10   In the last decades of the century, becoming the first destination in quantitative terms.

11   See the 'Introduction' in this volume.

public sector; and, finally, a relatively weak 'brand image', unable to capture the masses crowding other Mediterranean resorts.

The concrete objectives of the new plan were quite ambitious:[12]

- 10 million tourists by 2010 (7 from abroad);
- 160,000 new bed spaces, bringing overall capacity up to 230,000;
- total investments to reach 9 billion Euros;
- earnings of 48 billion Euros per year;
- 600,000 new jobs;
- tourism to increase its contribution to the GDP to about 20 percent (compared to 8.5 percent in 2000)

In order to achieve these objectives, the Government aimed to diversify tourist products, realize new structures and infrastructures, provide training for operators, with a new focus on marketing and promotional strategies, and the valorization of the natural and cultural 'resources' of the country. Beyond these quite common-place strategic guidelines, however, King Mohammed VI (who chaired the *Assise du Tourisme*), called for a radically new way of looking at (and living with) tourism: tourism, in his words, was not simply about selling an exotic landscape and experiences of *dépaysement* and adventure but, rather, tourism:

> beyond constituting an economic activity of great importance, is also the culture – and the art – of communicating with the Other. In this perspective, its development requires a careful management of our natural resources [...] but also the heritage of our civilization and millennial culture, known for its tradition of hospitality (cited in El Amrani 2001: 26).

In this understanding, mass tourists should not be simply conceived as 'spectators of the vestiges of Moroccan civilization', but rather the witnesses of the dynamic history of a millennial culture: '[...] it is the duty of every Moroccan to consider themselves a tourist entrepreneur, in order [for us] to succeed in this endeavour. [We must] create a new welcoming culture for tourists, since we are the 'hosts' in our country [...]' (ibid., 27).

The King's invocation, in many ways, became a launching pad for some of the most innovative initiatives of the years that followed, while also profoundly shaping the ways in which the country began to represent itself to the new masses eager to approach (and pay for) an 'accessible' and 'hospitable' Morocco. Marrakech, in particular, became a key focus of many of these initiatives.

---

12   See http://www.tourisme.gov.ma.

Figure 2.4   The Tea Ceremony

Source: © Brochure of the Office National Marocain du Tourisme.

New 'Cultural' Geographies of Tourism

The promotional stand of the *Office National Marocain du Tourisme* at the 2008 *Bourse Internationale du Tourisme* held in Milan, placed strong emphasis on two strands of mass tourism: first, on the 'Marrakech product', promoted with a CD entitled '*à la magie de Marrakech Medina*'; secondly, on what were previously considered niche products: golf and trekking in the High Atlas mountains (both presented only in English, signalling a new strategy for new markets). Overall, the 'Moroccan experience' was summarized in 5 themed chapters: 'Beach & Relaxation'; 'Culture & Discovery'; 'Sensations & Well-Being'; 'Golf & Art of Living' and 'Business & Relaxation'.[13] The new brochure tellingly shows an intriguing compromise between some traditional stereotypes of Morocco and the search for a more sophisticated visual and textual language that transcends some of the 'banalities' previously utilized to address mass tourism.

The cover of the catalogue portrays the traditional Moroccan mint tea, served on a tray covered in rose petals, offered up to the reader (Figure 2.4). It is with this 'welcoming' gesture that the travel experience of Morocco begins. The brief textual introduction is in line with the spirit of the images:

13   See http://www.tourisme-marocain.com.

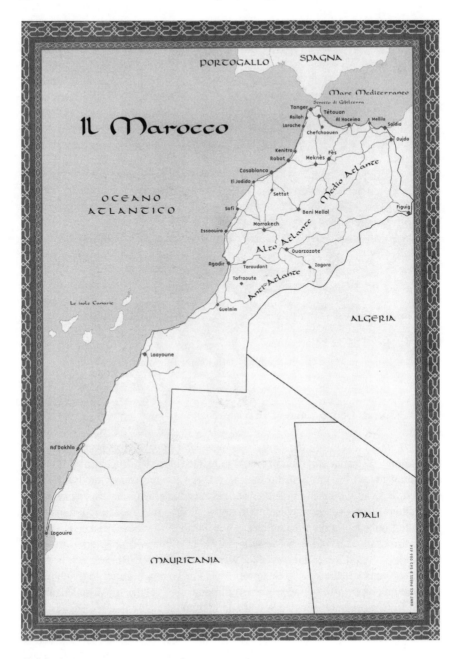

Figure 2.5   An 'orientalized' map of Morocco

Source: © Brochure of the Office National Marocain du Tourisme.

Located between the Mediterranean, the Atlantic Ocean and the Sahara desert, Morocco is a country where Nature let its imagination run free. It is a land of stunning contrasts, characterized by a culture and an atmosphere that bring together, in the Moroccan Kingdom, the most vivid dreams and Oriental traditions. From the peaks of the Rif and the Atlas Mountains covered in snow, to the verdant oases of the desert, from its golden beaches to its luxurious forests, from its Imperial cities to its Berber villages and Casbahs, Morocco offers the traveller a fabulous odyssey, a trip into Wonderland.

The catalogue is organized, moreover, according to a newly conceived cultural geography of tourism (see Figure 2.5).

'Marvelous travels into the North of Morocco' is, significantly, the focus of the first section, highlighting cities without a long standing tradition of mass tourism,[14] but that today are at the core of the future geographies of tourism envisioned by the Government: Tangiers, Tetuan, Chefchaouen, Asilah and Larache. This part of the country is seen as central to the development not only of new forms of domestic tourism, but also of international beach tourism, albeit with a pinch of culture and exoticism.

Figure 2.6   European tourists enjoy a panoramic view from above Fès
Source: © Brochure of the Office National Marocain du Tourisme.

---

14   With the partial exception of Tangiers in the first decades after the war.

The second section, entitled 'Marvelous travels into the heart of Morocco' is focused, instead, on Fès, Meknès, Salé, Rabat, Casablanca, Safi and El Jadida. The 'heart' metaphor evokes a presumed geographical core, but also the political and historical 'centre' of the country, since the list includes three out of the four so-called 'Imperial Cities'. The opening pages depict a couple overlooking a vast urban area (see Figure 2.6), occupying a privileged standpoint that allows them to 'take-in' the landscape that lies below. The chosen spot is, in fact, one where all the guides in Fès take their tourists for a 'proper' visual experience of the most celebrated medina in the world. Tellingly, the couple is dressed in white linen, reminiscent of a colonial experience of the Orient.

The third section is dedicated instead to 'Marvelous travels into the Moroccan South'. Here, the Orientalist aesthetic is taken to its extremes in describing Marrakech, Essaouira, Agadir, Tafraoute, Taroudant, Ouarzazate and Zagora. Agadir, previously promoted as a stand-alone beach destination, is now associated with other more 'cultural' destinations, in line with the new intention of the Government to convert conventional (beach) mass tourism – where Morocco still suffers fierce competition from other Mediterranean destinations – into cultural mass tourism. The catalogue, in fact, significantly omits the word 'tourist', replacing it with the term 'traveller', more apt to convey the 'cultural' nature of the new travel experience of Morocco.

The last section is even more innovative, highlighting 'Marvelous travels into the North-East of Morocco', from Al Hoceima to Melilla, Oujda, Saidia and Figuig. This is clearly an attempt to expand and diversify the geographical reach of a form of mass tourism that is rapidly entering a mature phase in some destinations like Marrakech. Here, some of the traditional stereotypes (such as the '*á la menthe*' depicted on the cover – Figure 2.4) are now accompanied by new products and a 'cultural twist' with new itineraries and new themes.

This re-crafting of mass tourism is confirmed in the closing pages of the catalogue, dedicated to a description of the cultural features of Morocco, a country 'ready to open its arms to the tourist'. Some of the major themes/resources highlighted are:

'Hospitality'

> Encounters with the Moroccan people are the most important of the Kingdom's riches, where hospitality is still firmly in place in all of its provinces. The Moroccan population is of Arab, Andalusian, Berber and African origins, each characterized by diverse traditions and ways of life; traditions waiting to be discovered.

Here, the colonial description of different 'types' of Moroccans – a sort of *Musée de l'Homme vivant* – is reminiscent of the cultural typologies described by French geographer George Hardy in his influential *Le Maroc* (1930). Just like Hardy's imagined geographies, the '*découverte du Maroc*' promised by the catalogue englobes both fascinating landscapes and mysterious places, but also 'curious' people, and radically-different, objectified Others: 'a diversity evident in the various dialects that characterize the country's population, as well as their different modes of dress'. This mélange of colourful faces and peoples, however, is presented as part of a single mosaic making up the nation – but also the 'Morocco brand', offered up for the consumption of the masses.

'The Souk'

> It is an endless feast for the eyes, a cocktail of spices and other enticing smells, an intricate web of alleyways that never seem to end. Every Moroccan city has its own Medina, with its noisy crowds, enterprizing salesmen, shops submerged in an Oriental *chiaroscuro*: ordinary scenes of an extraordinary life.

Here, too, the influence of Orientalist literature persists – although the 'web of alleyways' and the 'Oriental *chiaroscuro*' do not evoke a dangerous, potentially rebellious Orient (and Orientals) lurking in the medina, but rather a fascinating, dusty-orange colonial tableau, where poverty and backwardness are presented (and arguably perceived by the tourist) as markers of tradition and cultural 'purity'. The 'noisy crowds' and the 'enterprising salesmen', become reassuring caricatures performing the *souk* for the Western gaze.

'The Sahara'

> Its history is that of Humanity itself. A mineral universe that gave birth to a whole continent, harbouring within it countless hidden treasures. An immensity of sand and rocks, of oases and mirages, the Sahara remains still today the land of enchantment and dreams of escape.

This is arguably the most innovative element of the presentation, although the language remains colonial, recalling an age of exploration and enchantment with the desert (see Rondeau 2000). This description evokes popular (and postcolonial) European understandings of the desert as an 'empty' space whose immensity inspires (in the contemporary traveller) meditation and self-reflection (in many ways echoing the nineteenth century trope of the lone (male and white) traveller-hero). But there is also an important new geopolitical dimension:

> At the gates of the desert, along the Atlantic shore, several sites draw the traveller's attention: Tan-Tan Beach and further down the coast, Cape Juby, then Tarfaya… And Laayoune, emerging from the sand in its triumphant modernity. But we should also not forget Smara, the spiritual capital of the Sahara, in the heart of the Saguia al Hamra. And in the extreme south, at the sources of the Rio de Oro, Dakhla with its port known for sports fishing, and the most beautiful bay of the southern Sahara.

Here, the former Western Sahara (annexed by Morocco in 1975 and still the object of fierce political controversy) is presented simply as a semi-pristine tourist destination.

These new tourism geographies reveal a concerted attempt to re-write Morocco's imaginary and real places through the Orientalist eye – but also through the mundane, rather banal expectations of a hypothetical (again, standardized and Orientalized) European tourist. Nonetheless, the re-staging of the colonial for the masses is not a linear process where docile travellers simply buy into official rhetoric, as though visiting a fascinating *Musée Vivant* populated by equally docile 'locals'. On the contrary, our thesis is that the 'cultural turn' in Moroccan tourism, when translated (in place) into mundane practices, often produces new spaces where meaning is endlessly negotiated and where stereotyped representations and performances of colonial aesthetics are challenged and sometimes even disrupted by both the tourists and those who work with/for them.

L'appel de Marrakech

In 2003, two years after the launch of 'Vision 2010', Marrakech was already leading the growth of mass tourism in Morocco with about 30 percent of the foreign tourist revenues, and more than a quarter of total arrivals (Moujahid 2004: 1). According to a report released by the *Conseil Régional du Tourisme*, in 2004 the number of tourists in Marrakech went up by another 25 percent compared to the previous year.[15] This trend continued in the following years (*L'Economiste* 1 April 2005): in 2007, Marrakech registered a record number of presences (almost 50 percent of the national figure), also thanks to a campaign of investments that significantly increased the city's hotel capacity. The new international airport – second only to Casablanca in terms of air traffic – is now a key international gateway to the country, especially with the launch of new low cost connections with many European cities. *development of*

Although these figures are seen by the authorities with a degree of optimism, *Tourism's* the final objective remains that of hosting 3.4 million tourists by 2010, in line with the national plan. By 2010, Marrakech expects to open 22 new hotels, 30 new tourist residences and seven tourist villages, including a second *Club Méditerranée* (Moujahid 2004: 1). In order to achieve these ambitious goals, in 2004 the Government launched a comprehensive project of urban renewal, whose main objective is the reorganization of some key urban spaces in Marrakech according to the 'cultural and touristic vocation of the city' (Borghi 2008a, 2008b). Marrakech is, in fact, increasingly seen as the place where the whole country is on 'display'; where all of the diverse elements that compose today's Moroccan identity come together. Marrakech is presented to the tourists as the place where they can encounter the materialization – that is, the confirmation – of a long series

---

15   See http://www.emarrakech.info.

of Orientalist tropes about Morocco. Here, Moroccan 'culture' can be staged for the masses, can become reachable and manageable. The whole city, in its role of tourist capital, is seen as the Moroccan 'product' par excellence, an international brand-image for the entire country.

According to many critics, the satisfaction of the tourist gaze has become an absolute priority in this new grand plan of tourist development,[16] a priority that in practical terms translates many of the city's spaces into a sort of living museum feeding off global images of 'the seduction of Marrakech'. Here, the tourist rhetoric, in its attempt to restage the colonial for the European masses, encounters the materiality of the city and the 'real' places where tourism is practised and performed every day. This also explains the recent emphasis of the authorities on Morocco's 'living heritage', presented as a unique patrimony to be preserved (Filali 2008).

In Marrakech itself, it is an entire lifestyle that must be protected (El Faiz 2002) – and possibly put on display. The promotion of the 'Marrakech product' is thus based on the assumption that 'here', real contact with the local people and real access to Moroccan culture is actually possible. Marrakech is presented as a city redolent of history, rich in monumental material witnesses of the country's grand past. Its world-famous square, the Jamaa el Fna becomes a metaphorical urban icon, a bridge between the past and the present, the place where (spectacularized) Moroccan tradition encounters modernity (for example the tourists and Morocco's new global role):

> Eternal as the snows on the highest peaks, imposing as the mountains of the High Atlas, rooted in history as the palms that fill its oases, Marrakech is the final touch in an image of infinite beauty.
> The greatest kings have fought over it, countless noble dynasties have governed it, innumerable sages, artisans, architects, painters and sculptors have filled and embellished its stunning palaces, mosques, gardens and medersas ...
> Marrakech: the Imperial city that gave its name to Morocco
> Here, Berbers and Arabs come together, nomads and mountain dwellers descend to sell their crafts, artisans flourish, and it is the paradise of merchants.
> Together with this, countless palaces, hotels, restaurants, golf courses, casinos: Marrakech, the capital of the Moroccan South.
> For its thousands of riches accumulated in thousands of years, for the enchantment of the senses, don't miss Marrakech (Brochure, ENMT).

Interestingly, Marrakech is promoted through the very same images and slogans for both the European and Moroccan readership. The Government's campaign, in fact, has the double aim of improving the country's image abroad, but also identifying a key locus where Moroccans themselves can see the concrete and

---

16   Although it was already in Lyautey's times, as the Protectorate's architects re-
structured entire city quarters.

same image abroad + nationally

Figure 2.7   Jamaa el Fna at sunset (photo by Claudio Minca)

(internationally) celebrated signs of their glorious past and fascinating present culture of the country (Filali 2008). In such depictions, Marrakech is presented as the heart of the country, with the Jamaa el Fna seen as the most representative place for experiencing Moroccan cultural identity (Figure 2.7).

Marrakech, in fact, is now (as in the colonial days) also promoted as a destination for business tourism in search of sophisticated exoticism and top quality hospitality:

> Only three hours of flight from key European cities, connected by both charters and regular flights, Marrakech is one of the favourite destinations of those who desire strong emotions […] The investments of the past decade have made it the star of the Kingdom, not only the second home for countless international artists, businessmen and celebrities, but also a key destination for the international jet-set – and one of the favoured destinations for business tourism (Brochure, Conseil Régional du Tourisme de Marrakech 2008).

Marrakech is sold as the ultimate cosmopolitan outpost for those in the know:

> Marrakech, an evasion in time and space. Marrakech is, without a doubt, the 'in-city' of Morocco, with its hotels, luxurious riads, nightclubs and innumerable events […] A unique *cachet*, a 'garden city', a perfect combination between tradition and modernity (ibid.).

Needless to say, this marketing strategy and its material consequences (including the rapid growth of mass tourism) have been the object of controversy. Moroccan writer Abdelhak Serhane has strongly criticized such representations of the 'Red City', going as far as to compare it to a prostitute, plying 'easy exoticism' to passing tourists:

> They have dressed her and made her up like a hooker [...] The tourists, greedy to claim her, do so in an almost mechanical delirium. The city of Kings, now reduced to a vulgar postcard [...]. Marrakech el Bahja [Royal Marrakech] has become Marrakech el Hamra [Red Marrakech]. Red with shame, made-up with the powder of lies and artifice [...] Jamaa el Fna, made-up with heavy black eyeliner and red lipstick, gets inebriated every night to the sounds of music brought back from the grave, and wakes up every morning a destroyed, disarrayed Madame, with her wrinkles and failed dreams (Serhane 1999: 68–9).

Serhane's evokes, in many ways, the emphasis given in the crafting of the new Morocco 'product' to the *éblouissement de sens* – the 'enchantment of the senses' so often evoked in the tourist literature – and its heavily Orientalized (and often feminized) representations.

In the next section, we try to highlight some of the material effects of the new plan on the tourist colonization of the medina in Marrakech. The medina has become, in many ways, the new frontier for the restaging of the colonial for the masses, and it is here that the ambivalent relationship between images and practices is most evident – but also the object of contestation/negotiation.

The Marrakech Effect

> [...] The Moroccan tea ceremony is a privileged moment of the day. sharing a tea with Moroccans gathered around their 'berrad' (teapot) is an indispensable social ritual – and the best way of entering into contact with the culture of an always hospitable people (ENMT Catalogue 2008: 27).

Returning one night following a dinner in Jamaa el Fna to the Riad[17] owned by a Venetian friend of ours, located in the heart of the Marrakech medina, we are greeted by the beat of African drums and chanting, accompanied by applause and loud explosions of laughter. From the roof-top terrace of our Riad – once a space dedicated exclusively to women's domestic activities – we soon see what the whole fuss was about: on the terrace of a nearby Riad, a group of newly-arrived Northern

---

17   The Riads are traditional homes located in the medina (see Figure 2.8). Most have an internal courtyard with a series of adjoining rooms used as bedrooms or sitting rooms. They are usually disposed on two floors and have a roof-top terrace. For a fuller description, see Wilbaux (2001).

Figure 2.8   A riad in Marrakech (photo by Rachele Borghi)

Italian tourists are being 'welcomed' with a traditional *gnawa*[18] show and an induction of sorts to Morocco, Marrakech and its medina. The performance begins with the 'tea ceremony' that, as the guide explains to the newcomers, 'represents the very essence of Moroccan life and the symbol of Moroccan hospitality'. The Italian women are asked to put on turbans; the men are invited to accompany the *gnawa* performers's drums with rhythmic hand clapping. In a crescendo of sound and excitement, the local guide (dressed in the traditional garb of desert Tuaregs) 'explains' Morocco, 'land of traditions and culture', where, tea, music and dance are 'the heart of social life'. The Riad and the Medina are ideal sites, he tells the visitors, from which they can 'penetrate' an urban culture that 'has found a balanced compromise between modernity and a life style that goes back to millenarian traditions'.

This is nothing particularly new, of course, for the guide's presentation is a faithful reflection of some of the tropes that have marked official tourist rhetoric

---

18   The *gnawa* are seen as a distinct ethnic group, descending from Black African slaves. The music of the *gnawa* ceremonies is fast and rhythmic and used to induce trance-like states. In recent years, some gnawa performers have achieved international acclaim among 'world music' fans, thanks in large part to a Festival that takes place in Essaouira every summer.

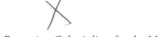

for over a decade now. What is interesting here, however, is that the tourists, despite appearing to be completely 'taken' by the atmosphere of the 'welcoming ceremony', begin asking questions that are too banal to be taken seriously: 'where does this tradition of good taste and exoticism come from?'; 'what is the typical local dish?'; 'should we eat with our hands as the Moroccans do?' and so forth. But even more interestingly, the visitors comment some of the staged scenes in their local dialect (since the guide presumably understands Italian): 'this tea is too sweet, how can they drink it?'; 'that dancer reminds me of my Moroccan employees back in Italy'; but also, 'let's be quiet and pretend that we are interested, otherwise the guide will take offence...'. The show continues for several hours, with the bored performances of the *gnawa* and the increasingly blasé comments of the guide accompanied by sexist jokes on the part of the tourists (for example, about their wives's interest for the dancers). Suddenly, a mobile phone rings: it is a friend from Italy, who receives a full (caricatured) description of the scene.

Now, the most intriguing aspect of this parade of banalities (and blatant vulgarity) is that it is repeated at least twice a week. The tourists change, their 'transgressive' comments might be different according to their background and mood, and to the chemistry of the moment. But the overall performance, as well as the attempts to implicitly disrupt the meaning of the 'representation' (Goffman 1959) on the part of the tourists – and often on the part of the dancers and the guide as well – is a sort of common pattern that emerged in almost all the performances we witnessed. Twice a week, the tourists are told the same stories (and the same jokes) and are welcomed with the same music and choreography. The extreme standardization of the tea ceremonial and the repetitive – and therefore paradoxical – nature of the whole performance are in fact partially 'resisted' by both the local operators and the tourists. These latter, with their (not always) subtle comments and jokes, in many ways banalize (and thus disrupt) the *willing suspension of disbelief* that the travel experience is supposed to produce (see Minca and Oakes 2006), and that the tourist rhetoric about Morocco promises. In other words, the meaning of the performances of 'immemorial Moroccan culture and hospitality' on the rooftop of the Riad is constantly negotiated and partially disrupted by another equally important set of performances: those enacted by the disenchanted comments of both the tourists *and* the local performers (each commenting the scenes in their 'local' language). We can see such negotiation and disruption as forms of re-appropriation and re-interpretation of the 'philosophy of the thing' (Mitchell 1988; but also MacCannell 1999, 2001; Lorimer and Lund 2003) on the part of both tourists and 'locals'; something that reveals the somewhat mundane nature of those same practices, and that reflects a pattern of cultural exchange that seems to be well known by all the protagonists in their respective roles – beyond (and despite) the 'magic' of the colonial re-enactment of Moroccan culture.

Two elements are of particular relevance here: firstly, while in most tourist representations of 'local culture', repetition and banality are paramount and are presumably often disrupted by the subjects involved, at the same time this disruption must remain constrained within the limits that allow the whole thing to maintain some degree of credibility (see Goffman 1959). This threshold of 'make believe' is a rather interesting grey area that would be worthwhile exploring in depth, but that is beyond the scope of this chapter (see Rojek 1996). The second element has to do with the 'real' consequences of these performances. For example, the growing proliferation of Riads converted into hotels or bed and breakfast is having enormous effects on the urban fabric of the medinas of Marrakech and of other Moroccan cities (Kurzac-Souali 2006). This proliferation is the result of the growing interest on the part of European tourists for staying within the old city itself and being able to experience the atmosphere of a traditional Arab street (Escher et al. 2001a, 2001b, 2004). In Marrakech, the so called *maisons d'hôtes* (as some converted Riads are called – see Chebbak 2004, Saigh Bousta 2004a), have mushroomed, spurred by the European fascination for the 'magic' of the medina:

> Riads have brought back an art of living that the Moroccan people themselves have half forgotten. Windowless walls, in featureless side-streets conceal paradises to be discovered with over-growing wonder, leafy patios, fountains, birdsong, terraces where life is for the living and which offer stunning views of the medina and its minarets and, in the distance, the snowy peaks of the Atlas mountains (http://www.ilove-marrakesh.com/lesriads).

With their location in the heart of the medina, Riads offer the new 'cultural' tourists a luxurious, 'comprehensive experience of Moroccan life' – and access to spaces that are normally impenetrable: 'Riads evoke the idea of a secret world ; they allow the foreigner that inhabits them to feel the thrill of discovering [...] and perhaps even experiencing a small part of this magical, enigmatic place, the universe of a Thousand and One Nights' (Saigh Bousta 2004b: 159).

The partial gentrification of the medinas of some of the main Moroccan cities (Tangiers, Rabat, Fes, Essaouira) reflects a tendency towards the rapid colonization of traditional urban spaces on the part of a new and powerful colonial aesthetic that is very often at the origin of much controversy, but that appears, at present at least, irresistible (Chebbak 2004, De Graincourt and Duboy 2002). With few exceptions, the Riads are designed and decorated in order to (re)create an Oriental ambiance and a sumptuous lifestyle, often in sharp contrast with the poverty and the decay of the surrounding neighbourhood. Escher and Patermann (2001) suggest, indeed, that many foreigners staying in Riads enjoy the sophisticated 'colonial atmosphere' that emanates from every single detail of the hospitality provided. The 'aristocratization' of some protected spots in the medina and the development of cultural tourism focused on '*l'art de vivre marocain*' in some of the most deprived parts of the city are, unsurprisingly, the subject of much local controversy (see Kurzac-Souali 2007a, 2007b). The tourist authorities are often accused of encouraging the gentrification and the progressive repossession of what is seen as a fundamental part of Moroccan urban patrimony – and a heritage to be protected from the impact of tourism.

The restaging of the colonial for the masses is therefore fast becoming a highly relevant issue, here as in other parts of Morocco, reflecting in many ways a more general trend toward the aesthetization of the tourist experience. It also reveals, however, a parallel banalization of the cultural practices produced by the attempted massification of 'cultural' tourism – again, something particularly evident in Marrakech.

In the past few years, a growing literature (mainly in French and German) has analyzed the effects of the new fashion for Riads and its consequences on the local population. On the one hand, the purchase of Riads on the part of foreign investors is seen as an important factor in the rehabilitation of the built environment, having had some positive effects in terms of job creation and the survival of traditional craftsmanship, although further important measures need to be taken if the patrimony of the medina is to be protected.[19] Some, however, are less positive about the impacts of this new trend. Boudjafad (2005), for one, argues that this is a 'market' whose economic effects are often far-removed from the city, with accommodation normally paid for in advance – and often to foreign intermediaries. Marrakech-based scholar Kurzac-Souali (2005a, 2005b, 2006, 2007a, 2007b, 2007c), who has extensively studied this phenomenon, has highlighted how the emergence of new urban dynamics, coupled with the modification of existing urban cultures, has produced distinct ways of inhabiting the spaces of the medina: that of the autochthones, that of the new European residents, and that of the tourists (although these last two categories are very often confused by the local population – see Saigh Bousta 2004a). These different ways of inhabiting and experiencing the spaces of the medina have also been accompanied, however, by new processes of spatial fragmentation and residential segregation:

> The new inhabitants of the medina have transformed the built environment of certain neighbourhoods. Their arrival has also transformed, however, the social composition of some areas [with the emergence of] cosmopolitan quarters highly unequal in terms of income. [...] Although there appears to be a new social "mix" in the medina, it is a transitory, ephemeral one, hiding significant spatial segregation whose long-term effects must be taken into consideration [...] The "requalification" of old neighbourhoods has reinforced, in fact, socio-spatial and residential segregation [...] encouraging the emergence of a 'two-speed' medina (Kurzac-Souali 2005a)

What is fast becoming evident in the case of Marrakech is the emergence of a highly fragmented space, where gentrified areas of 'tourist value' rub shoulders with densely populated poor quarters (mainly in the north and the east of the medina) (Borghi et al. 2007). As Benzine (2000) notes, what emerges here is a

---

19   See the recent debates on *L'économiste* and *La Vie Touristique*.

conception of urban management that increasingly reinforces the dichotomy between economically viable spaces/heritage – and those that are 'useless', or at least not easily marketable.

Recent research by Saigh Bousta (2004b) reveals, nonetheless, that a sizable proportion of residents in the medina believe that the presence of foreigners brings positive effects, at least in terms of the quality of the urban environment and the creation of new jobs. Her study also highlights, however, concerns about skyrocketing house prices, lack of regulation (or at least respect of existing regulation) in renovation, but also the sexual behaviour of some tourists staying in the Riads. These concerns are unsurprising, given the rapid changes in the urban fabric. What is interesting, however, is that the most virulent condemnations of this phenomenon actually come from *outside* the medina. According to Kurzac-Souali (2007a: 87), the most significant criticism towards the progressive 'Westernization' of the Medina has come from the national media and, more generally, from middle class Moroccans that in most cases no longer live in the medina themselves, but are concerned about the loss of a key site for the reproduction and preservation of a 'truly Moroccan' urban tradition. We do not have the space here to comment on this important debate, or its protagonists. What must be noted, however (following Kurzac-Souali – but also the work of Saigh Bousta), is that we cannot take for granted that the tourist 'colonization' of the medina is perceived only as a threat – nor can we presume to speak for the medina's residents without some robust enquiry into the impact of tourist practices (in place).

Reasserting the Re-staging

According to Saigh Bousta (2004b), 57 percent of the residents interviewed declared that the gentrification of the medina was a 'good thing' for the urban environment as a whole, with 47 percent noting that it helped improve the 'tranquility' of their neighbourhood, 27 percent that it improved its 'safety', and 20 percent that it was important for the preservation of its 'architectural patrimony'. These survey results do not, of course, sum up all of the effects of the restaging of the colonial in Marrakech. What they hint at, however, is something that we believe is particularly important for the central argument of this chapter: the new forms of mass cultural tourism that are becoming increasingly important in Marrakech and elsewhere in Morocco represent a highly complex phenomenon, and cannot be analysed simply through a deconstruction of the neo-colonial imaginary they rely upon, or an a priori rejection of whatever social and cultural changes mass tourism generates. Existing studies and our own fieldwork in the medina of Marrakech show, if anything, that a different type of scrutiny is required. Such scrutiny needs to engage, case by case, with both the representations that help craft the material restaging of the colonial for the masses – and the tourist practices linked to those very representations (on this see Franklin and Crang 2001). It also must engage, however, with the ways in which such representations and practices are perceived and re-staged by the residents and by those who work for and with the tourists. This chapter, therefore, does not claim to provide any broad generalizations regarding the impact of the restaging of the colonial in contemporary Morocco. We have simply tried to critically engage with a discursive formation (and a specific set of practices) that has gained extraordinary currency on the Moroccan cultural and political scene, and that reveals the extraordinary resilience of the colonial aesthetic. The colonial restaging for the masses, nonetheless, is subject to endless re-interpretation that translates, quite literally, into ever new geographies. What we hope that this chapter has helped highlight is that such geographies are, on the one hand, the outcome of the practical enactments of the discursive formation that we have named here as '*L'appel du Maroc*' on the part of the Moroccan authorities and European tour operators, in order to produce and structure 'real and imagined' tourist spatialities. On the other hand, however, such geographies are also the expression of a series of practices on the part of the tourists and the residents who come into contact with them: it is these endless encounters, in place, that give life to a fascinating (and only partially explored) set of interpretations and performances, within which the official tourist rhetoric is at the same time confirmed and transgressed. It is in specific sites, we argue, that representations and practices merge; and it is there, we would like to suggest, that we have to 'redirect our gaze' in order grasp the 'deeper meaning' of the real and imagined tourist experience of Morocco.

References

Bærenholdt, J.O., Hadrup M., Larsen J., and Urry, J. 2004. *Performing Tourist Places*. Aldershot: Ashgate.

Barthou, L. 1930. *Lyautey et le Maroc*. Paris: Petit Parisien.

Bauchet, J. 1954. Le Casinò. *Perspectives d'Outre Mer: Notre Maroc*, 1(13), 43–4.

Bélanger, C.E. and Sarrasin, B. 1996. *Développement et Tourisme au Maroc*. Paris: L'Harmattan.

Benoist-Mechin, J. 2007. *Lyautey l'Africain, ou le Rêve Immolé (1854–1934)*. Paris: Perrin.

Benzine, H. 2000. *Schéma Directeur d'Assainissement Solide du Grand Marrakech. Les Marrakechis face à leurs déchets*, PhD Thesis, DESS, Université Muhammed V, Rabat.

Berriane, M. 2003. Bilan sur le tourisme marocain. *Proceedings of the Rencontre internationale de Fes 'Patrimoine et development durable des centres historiques urbains'*. Rabat: UNESCO.

Birkeland I. 2005. *Making Place, Making Self*. Aldershot: Ashgate.

Bissell W. C. 2005. Engaging colonial nostalgia. *Cultural Anthropology*, 20(2), 215–48.

Boëhm, B. 1993. Arts et séductions: synthèse, in *Images et colonies*, edited by P. Blanchard and A. Chatelier. Paris: Syros-Achac.

Borghi, R. 2004. Dove finisce l'altra sponda? Tra Mediterraneo e 'marocchinità' attraverso il turismo culturale di Marrakech, in *Orizzonte Mediterraneo*, edited by C. Minca. Padova: Cedam.

Borghi, R. 2008a. *Geografia, Postcolonialismo e Costruzione delle Identità: Una lettura dello Spazio Urbano di Marrakech*. Milan: Unicopli.

Borghi, R. 2008b. La mise en scène de la ville: regard sur l'espace urbain de Marrakech coloniale et postcoloniale, in *Le fait colonial au Maghrèb,* edited by N. Maarouf. Paris: L'Harmattan.

Borghi, R. and Minca C. 2003. Le lieu, la place, l'imaginaire: discours colonial et littérature dans la description de la Jamaa el Fna, Marrakech. *Expressions Maghrebins*, 1(2), 155–74.

Borghi, R., Lando F. and Senn M. 2007. Aménagement touristique et transformation de l'espace urbaine: les risques du développement du secteur à travers le cas comparé de Venise et Marrakech', in *Aménagement du territoire et risques environnementaux au Maroc*, edited by B. Akdim and M. Laaouane. Fès: Publications de l'Université Sidi Mohamed Ben Abdellah.

Boudjafad, S. 2005. Descente dans les maisons d'hôtes. *Aujourd'hui Le Maroc*, 24 March.

Cattedra, R. 1990. Nascita e primi sviluppi di una città coloniale: Casablanca, 1907–1930. *Storia Urbana,* 53, 127–79.

Chebbak, M. 2004. Maisons d'hôtes : un avar orientaliste. *Architecture du Maroc*, 17, 29–30.

Clément, J.F. 1994. Lyautey à Marrakech. *Horizons Maghrebins*, 23/24, 15–22.

Coleman, S. and Crang, M. 2002a. *Tourism: Between Place and Performance.* New York: Berghahn Books.

Compagnie des Chemins de Fer au Maroc. 1921. *Hotels et gites d'étape.* Rabat: Africaines.

Crang, M. 1999. Knowing, tourism and practices of vision, in *Leisure/Tourism Geographies*, edited by D. Crouch. London: Routledge, 238–57.

Crang, M. 2003, Qualitative methods: touchy, feely, look-see? *Progress in Human Geography*, 27(4), 494–504.

Crang, M. 2004. Cultural geographies of tourism, in *Companion of Tourism Geography*, edited by A. Lew, M. Hall and A. Williams. Oxford: Blackwell, 74–84.

Crang, M. 2006. Circulation and emplacement: the hollowed out performance of tourism, in *Travels in Paradox: Remapping Tourism*, edited by C. Minca and T. Oakes. Boulder, CO: Rowman & Littlefield, 47–64.

Crouch, D. 2004. Tourist Practices and Performances, in *Companion of Tourism Geography*, edited by A. Lew, M. Hall and A. Williams. Oxford: Blackwell, 85–96.

Crouch, D. 2005. Flirting with space: tourism geographies as sensuous/expressive practice, in *Seductions of Place*, edited by C. Cartier and A. Lew. London: Routledge, 23–34.

Daoud, A. 2001. Dans les coulisses des assises. *Le Temps du Maroc*, 272, 28–9.

De Amicis, E. 1913. *Marocco*. Milan: Fratelli Treves.

De Foucauld, C. 1883. *Reconnaissance au Maroc*. Paris: Challamel.

De Graincourt, M. and Duboy, A. 2002. Riads, l'irresistible attrait. *Medina,* 9, 71–84.

Delacroix, E. 1893. *Journal*. Paris: Plon.

Delacroix, E. 1999. *Souvenirs d'un voyage dans le Maroc*. Paris: Gallimard.

Département du Tourisme. 2001. *Le secteur touristique: Statistiques 2000*. Rabat: Direction de la Planification et de la Coordination de la Promotion.

Doury, P. 2002. *Lyautey*. Paris: L'Harmattan.

Dugard, H. 1918. *Le Maroc de 1918*. Paris: Payot.

Edensor, T. 2001. Performing tourism, staging tourism: (re)producing tourist space and practice. *Tourist Studies*, 1(1), 59–81.

Edensor, T. 2006. Sensing tourism spaces, in *Travels in Paradox: Remapping Tourism*, edited by C. Minca and T. Oakes. Boulder, CO: Rowman & Littlefield, 23–46.

El Amrani, N. 2001. L'art de communiquer avec l'autre. *Le Temp du Maroc*, 272, 26–7.

El Faiz, M. 2002. *Marrakech, Patrimoine en Péril*. Paris and Rabat: Acte Sud and Eddif.

Enjeux, A. 1993. Dossier tourisme, la reprise? *Le magazine de l'Entreprise et de l'Economie*, October, 57.

Escher, A., Petermann S. and Clos B. 2001a. Le bradage de la médina de Marrakech?, in *Le Maroc à vielle du troisième millénaire – Défis, changes et risques d'un développement durable*, edited by Berriane and Kagermeier. Rabat: Publications de la Faculté des Lettres et de Sciences Humaines de Rabat.

Escher, A., Petermann, S. and Clos B. 2001b. Gentrification in der Medina von Marrakech. *Geographische Rundschau*, 53(6), 24–31.

Escher, A.J. and Patermann S. 2001. Neo-colonialism or gentrification in the medina of Marrakech. *ISIM NEWSLETTER,* 5, 34.

Filali, J. 2000. Pour un environnement prophylactique. *Marrakech Informations*, 164, 10.

Franco, G. 1996. Le pays, in *Aimer le Maroc. Monde et voyages*, edited by J. Bonhomme-Penet and O. Dénommée. Paris: Larousse.

Franklin A. and Crang, M. 2001. The trouble with tourism and travel theory? *Tourism Studies*, 1(1), 5–22.

Galissot, R. 1999. *Le Tourisme au Maroc: Notes*. Casablanca: Direction des relations économiques extérieures.

Goffman, E. 1959. *The Presentation of Self in Everyday Life*. New York: Doubleday & Company.

Gregory D. 2001. Colonial nostalgia and cultures of travel: spaces of constructed visibility in Egypt, in *Consuming Tradition, Manufacturing Heritage*, edited by N. Al Sayyad. London: Routledge, 111–151.

Henderson C. and Weisgrau, M. 2007. *Raj Rhapsodies: Tourism, Heritage and the Seduction of History*. Aldershot: Ashgate.

Hillali, M. 2005. *La Politique du Tourisme au Maroc*. Paris: L'Harmattan.

Hillali, M. 2007. Du tourisme et de la geopolitique au Maghreb: le cas du Maroc. *Hérodote*, 127, 47–63.

Hoisington, W.A. 1995. *Lyautey and the French Conquest of Morocco*. New York: St.Martin's Press.

Kurzac-Souali, A.C. 2005a. La revalorization de la médina dans l'espace urbain au Maroc. Un espace urbain revisité par les élites et le tourisme, in *Villes réelles, villes projetées. Fabrication de la ville au Maghreb*, edited by Boumaza et al. Paris: Maisonneuve & Larose, 377–90.

Kurzac-Souali, A.C. 2005b. Ces riads qui venedent du reve, patrimonialisation et segregation en medina, in *Habiter le patrimoine. Enjeux, approches, vecu*, edited by in M. Gravari-Barbas. Rennes: Presses Universitaires de Rennes, 467–478.

Kurzac-Souali, A.C. 2006. *Les Medinas Maroccaines: une Requalification Selective. Elite, Patrimoine et Mondialisation du Maroc*, PhD thesis, University of Paris IV Sorbonne, Paris.

Kurzac-Souali, A.C. 2007a. Rumeurs et cohabitation en médina de Marrakech: l'étranger ou on ne l'attendait pas. *Hérodote*, 127, 64–88.

Kurzac-Souali, A.C. 2007b. La médina vue par ses nouveaux habitants: représentations et usages d'une citadinité retrouvée. *Le Cahiers d'Al Omrane*.

Kurzac-Souali, A.C. 2007c. Engouement mediatique et reconquete dea maisons traditionelles au Maroc. *Espace et Culture*.

Laprade, A. 1934. *Lyautey urbaniste. Souvenirs d'un Témoin*. Paris: Horizons de France.

Lorimer, H. and Lund, K. 2003. Performing facts: finding a way over Scotland's mountains, in *Nature Performed: Environment, Culture and Performance*, edited by B. Szerszynski and C. Waterton. Oxford: Blackwell, 130–44.

Loti, P. 1988 (first edition 1889). *Au Maroc*. Paris: La Boîte à Documents.

Lozato-Giotart, J.P. 2008. *Geographie du Tourisme: de l'Espace Consommé à l'Espace Maîtrisé*. Paris: Pearson Education.

Lozato-Giotart, J.P. 1991. *Le Maroc*. Paris: Karthala.

Lyautey, L.G.H. 1918. *Les Guides Bleus. Maroc*. Paris: Hachette.

Lyautey, L.G.H. 1995 (first edition 1927). *Paroles d'action*. Paris: Editions la Porte.

MacCannell, D. 1999. *The Tourist: A New Theory of the Leisure Class*. New York: Schoken.

MacCannell, D. 2001. Tourist agency. *Tourism Studies*, 1(1), 23–38.

Mahr, K. 2007. In colonial nostalgia a seed of democracy? *Time Magazine*, 10 May. Available at http://www.time.com/time/world/article/0,8599,1619291,00. html.

Maurois, A. 1931. *Lyautey*. Paris: Plon.

Minca, C. 2006. Re-inventing the 'square': postcolonial geographies and the tourist gaze in Jamaa el Fna, Marrakech, in *Travels in Paradox: Remapping Tourism*, edited by C. Minca and T. Oakes. Boulder, CO: Rowman & Littlefield, 155–84.

Minca, C. and Oakes, T. 2006. Introduction: travelling paradoxes, in *Travels in Paradox: Remapping Tourism*, edited by C. Minca and T. Oakes. Boulder, CO: Rowman & Littlefield.

Mitchell, T. 1988. *Colonizing Egypt*. Berkeley, CA: University of California Press.

Mitchell, W.J.T. 2005. *What do Pictures Want?* Chicago: Chicago University Press.

Morton, P. 2000. *Hybrid Modernities*. Cambridge: MIT Press.

Moujahid, M. 2004. Un fort potentiel touristique et une vocation agricole malgré tout. *La Vie Economique*, 26 March–1 April, 47.

Mouline, S. 1993. *Repères de la Mémoire. Marrakech*. Rabat: Ministère de l'Habitat.

Mourad, K. 1994. *Marrakech et la Mamounia*. Paris: ACR.

Oakes, T. 2005. Tourism and the modern subject: placing the encounter between tourist and other, in *Seductions of Place*, edited by C. Cartier and A. Lew. London: Routledge, 36–54.

Oakes, T. 2006. Get real! On being yourself and being a tourist', in *Travels in Paradox: Remapping Tourism*, edited by C. Minca and T. Oakes. Boulder, CO: Rowman & Littlefield, 229–50.

Peleggi, M. 2005. Consuming colonial nostalgia: The monumentalization of historic hotels in urban South-East Asia. *Asia Pacific Viewpoint*, 46(3), 255–65.

Potier, N. 2006. *Dix-Sept regards sur le Maroc*. Casablanca: EDDIF.

Rivet, D. 1980. Lyautey l'Africain. *Histoire*, December, 29, 16–24.

Rivet, D. 1984. Exotisme et 'penetration scientifique': l'effort de découverte du Maroc par les Français au début du XX siècle, in *Connaissance du Maghreb. Sciences Sociales et colonisation*, edited by J.C. Vatin. Paris: CNRS.

Rivet, D. 1988. *Lyautey et l'Institution du Protectorat Français au Maroc*. Paris: L'Harmattan.

Rivet, D. 2002. *Le Maghreb à l'Épreuve de la olonisation*. Paris: Hachette.

Rojek, C. 1996. *Decentring Leisure*. London: Sage.

Rondeau, D. 2000. *L'Appel du Maroc*. Paris: Flammarion.

Rosaldo, R. 1989. Imperialist nostalgia. *Representations*, 26, 107–22.

Saigh Bousta, R. 2004a. Le phénomène Ryad maison-d'hôte, esquisse d'une réflexion sur l'immersion culturelle du tourisme dans l'espace traditionnel des autochtones, in *Le tourisme durable*, edited by R. Saigh Bousta and F. Albertini. Marrakech: Faculté des Lettres et des Sciences Humaines.

Saigh Bousta, R. 2004b. Voisinage des Ryad maison-d'hôte dans la medina de Marrakech. Résultat d'une enquête réalisée en mars 2003, in *Communication Interculturelle, Patrimoine et Tourisme,*edited by R. Saigh Bousta. Marrakech: Faculté des Lettres et des Sciences Humaines.

Secréterait d'état au Plan, Royaume du Maroc Royame du Maroc, Haute Commissariat au plan 2005. *Point de Conjoncture de l'INAC*. Rabat: Institut National d'Analyse de la Conjoncture.

Turco, A. 1995. Delacroix in Marocco: indagine sull'altrove. *Terra d'Africa,* 4, 315–53.

Venier, P. 1997. *Lyautey avant Lyautey*. Paris: L'Harmattan.

Weisweiller, E. 1932. Marrakech, cité d'hivernage, in *Congrès International de l'Urbanisme aux Colonies et dans les pays tropicaux*, edited by R. Jean. La Charité-sur-Loire: Delayance.

Wharton, E. 2003. *In Morocco*. London: I.B. Tauris.

Wilbaux, Q. 2001. *La medina de Marrakech*. Paris: L'Harmattan.

Chapter 3
Banal Tourism? Between Cosmopolitanism and Orientalism

Michael Haldrup

Introduction: Routing Cosmopolitanism and Tourism

Does increased leisure travel and tourism lead to increased awareness of global connectedness and interdependencies or does it simply reinforce hegemonic orientalist scripts about Western superiority by transforming the globe into one big warehouse for the wealthy and (presumably) Western consumer? This is a normative concern that has haunted most theories of tourism from the beginning. In his history of vacation Orvar Löfgren captures this ambivalence of cosmopolitanism and orientalism in contemporary mass tourism by claiming that

> [V]acationing carries an emancipatory potential. At times new forms of mass tourism hold out the hope of changing the world, turning locals into cosmopolitans, breaking down artificial boundaries between nations, localities, classes, or generations – creating global communities. But moving out can also be a way of staying the same. While the skills of becoming a cosmopolitan can be important cultural capital for some, they don't work for others. Tourism can both open and close the mind (Löfgren 1999: 269–70).

While tourist theories generally have approached tourism in terms of consuming cultural difference the way tourist performances and experiences tie into people's everyday life has traditionally received less attention. In this article I show how tourist performances and experiences are an important part of the ways people position themselves in and as part of the world. Tourism is an important part of what has been called 'banal cosmopolitanism'. Thus, Beck has argued that the increased day-to-day experience of the global through consumption, media and so on produces a 'global cosmopolitanism! That whether it is reflexive or not transform the experiential spheres of life worlds enforcing a globalization of emotions and empathy' (Beck 2004: 151–2).

In continuation of this I will argue that tourist performances and experiences is a significant vehicle for the emergence of cosmopolitan orientations in everyday life. It is when we are performing tourism that we are most likely to achieve first-hand experience with other places, people and cultures and experience our connectedness with these. Thus, tourism may be viewed as a social, cultural and material field that enables and shape different modes of attachment to and detachment from places, cultures and people; different modes of belonging in the world; hence we may conceive of tourism as a way of positioning oneself in the world; of producing and reproducing the everyday geographies of 'banal cosmopolitanism'.

One of the reasons why these everyday geographies of tourist performances have received little attention is that tourism theory has been caught up in a representational that has left the mundane and banal aspects of tourism has often been left out of analytical sight. By emphasizing the importance of mundane banalities for the study of tourism performances, the purpose of this article is not to banalize tourism as has been done so often in social theory but to call attention to the 'banal' ways in which tourist performances produce, reproduce and make use of distinctions, relations and roles that tie into much broader politics and performances of everyday life. This significance of 'banalities' in tourism has especially been acknowledged by turn towards embodiment, materiality and performance in current writings on tourism. Whereas much earlier literature have been dominated by a 'representational' paradigm, focusing the analytical gaze on the symbolic and discursive aspects of tourism recent accounts of tourism in social and cultural studies have sought to dislocate attention from symbolic meanings and discourses (which dominated much earlier studies of tourism) to embodied, collaborative and technologized doings and enactments (Haldrup and Larsen 2006).

This turn towards theories of performance and performativity in tourist studies has in important ways challenged and contested fixed readings of tourist spaces by making ethnographies of what humans, institutions and non-humans do – enact and stage – in order to make tourism and performances happen. Tourism performances are surely choreographed by concrete guidance and cultural scripts, but tourists are not just written upon, they also enact and inscribe space with their own 'stories'. Like this, tourists carry quotidian habits and responses with them: they are part of their baggage and even the most 'exotic' tourist places afford stages for enacting the most 'intimate' and banal aspects of our life worlds (on this see Haldrup and Larsen 2003). However, tourist sites are also a staging ground for enacting the 'banal' identities and cultural stories are enacted. Rather than transcending the mundane, most forms of tourism are fashioned by culturally coded escape attempts. Not only do tourists bring their own bodies, their loved ones and their quotidian habits: they also bring the cultural stories in which their lives are immersed. Tourist places, such as the beach, the club, the promenade along the coast, are stages on which cultural stereotypes related to gender and ages are certainly enacted. Not least ethnic, national and regional stereotypes are continually being played out in the packaging and selling of 'local cultures' to tourists, as tourists bring preferences for particular imaginaries, settings, foods and so on with them. Billig (1995) coined the notion of 'banal nationalism', and tourist places are indeed places that are often penetrated by signs, symbols and set pieces affording banal nationalism to be enacted. Other tourists' ethnic and national belonging are eagerly analysed and compared with one's own, which literally may be flagged by way of national flags on top of sandcastles or handkerchiefs, bath towels and blankets in the national colours displayed on the beach or by the pool side (see also Löfgren 1999: 260 ff). Spectacular settings afford tourists with the performative spaces onto which his/her desires – for example playing with quasi-colonial kitsch – can be projected, and resorts offer familiar 'national' moorings in the 'exotic' sea, as when English pubs or Scandinavian restaurants are being designed as 'second homes' for the golden hordes from the North (Turner and Ash 1975). At the same time photographs, souvenirs purchased on vacation decorates family homes and in doing this 'colonizes' the times and spaces of non-tourist everyday life.

In this way, tourist performances and experiences feed into the more general fabric of mundane everyday life banalities. Objects, signs, narratives, foods, clothes that produce tourist places are themselves circulating just like the 'tourist' objects, images and so on that decorates houses, bodies, shelves and so at 'home. Such circulating objects and images connect the 'exotic' setting for tourist performances with the spheres of 'home'. However they may do so in unpredictable ways. The performative spaces of tourism are not only effects of encounters between 'naked' bodies and material landscapes, they are also inscribed with pre-existing cultural stories, memories, norms, fantasies, family networks, (post)colonial relations and commodity chains, and haunted by war and terror. These are not necessarily 'present' in place but they have effects in place. They circulate, they move and their meanings may change through a variety of local contexts when they are circulated across the globe as consumer goods or 'global news'. In doing this they may even act back on the viewer, on the tourist that encounters their effects in his holiday setting. An illustrative example of this was the publication of a series of cartoons of the prophet Muhammad in a Danish (regional) newspaper in autumn 2005 which travelled along diasporic ethnoscapes as well as the mediascapes of global news corporations causing local protests in several Muslim countries; riots that in turn produced local events such as the burning of Danish flags (West Bank, Palestine), buildings (Damascus, Syria) and calls for death penalties over the cartoonists and responsible editors (Pakistan) and through images, stories and newspaper headlines travel back along the same scapes and flows to the North European context and causing massive reductions in tourist travel to countries such as Turkey and Egypt. Some of the effects of this 'cartoon controversy' were that the number of Danish tourists booking a package tour to these countries halved, and the tourism industry blamed the newspaper, the cartoonist and the 'political climate' in Denmark. Thus the tourism industry in Turkey and other Muslim countries was affected negatively by a faraway event that was supposed to be a 'local' (yet distasteful) joke but ended up being 'global news'. Incidents such as this in which political events 'at home' circulate along unpredictable routes ending up acting back at the viewers from 'away' reminds us that tourism is a culture of circulation and connections. In the words of Appadurai such cultures 'transgress the boundaries of home and away, well-known and imagined, by creating specific irregularities because both viewers and images are in simultaneous in circulation. Neither images nor viewers fit into circuits of audiences that are easily bound within local, national or regional spaces' (Appadurai 1999: 4).

Moreover, such circuits also break down the barrier between the known and the fantasized. As Appadurai puts it: 'imagination is today a staging ground for action and not only for escape' (1999: 7). Like other cultures in motion, tourism is part of networks and circuits that are not easily located within national, local or regional spaces but encompass both localized performances in place as well as global processes. Indeed such circuits are constituent for the emergence of contemporary cultures of 'banal cosmopolitanism'. By being caught up in the circuits of such 'banal cosmopolitanism' tourism performances do not detach places and practices from the practices, social relations or cultural stories we 'normally' engage with. While the turn towards performance in the tourist literature surely has destabilized static and fixed conceptions of places and sites it has not sufficiently grasped the networked mobilities of objects, images, texts and technologies that permit tourism performances to take place and to be represented and (re)circulated across often great distances at various sites and times. While recognizing that localized tourism performances are framed by and draw upon global flows (of stories, objects, people, images, materials and so on), most ethnographies of tourism performances have not departed from the idea that 'ethnographies' take place within bounded sites. James Clifford (1997) famously argued that ethnography needed to leave behind its preoccupation with discovering the 'roots' of cultural and social forms and instead trace the 'routes' that produce and reproduce them. Taking inspiration from this I will argue that the 'banal cosmopolitanism' of tourist performances and experiences need to be 'routed' in order to grasp emerging cosmopolitan orientations in contemporary everyday life. This means that we need to trace out the routes that tie together the contexts 'at home' and 'away' in contemporary tourism and – especially how tourist performances tie into the ways we position ourselves as part of 'the global' – in order to reframe the ambivalence between cosmopolitanism and orientalism. Instead of asking if tourism and cosmopolitanism goes hand in hand, we need to understand what kinds of 'banal cosmopolitanism' are enacted when performing tourism.

In doing this I argue for the need to follow the routes that connect the contexts of 'home' and 'away' and therefore extending the ethnographic methods currently used in studies of tourist performances in particular tourist settings into the private home of tourist to examine how tourist performances are incorporated into the wider context of everyday lives. In this article I exemplify such modes of belonging in the world by tracing tourist performances and experiences from the front regions of tourist destinations and holiday making to the back regions in tourists private homes and lives. Thus, the examples discussed here are based on visits to and interviews in people's personal homes, and rests on tourist tales, souvenirs, and photographs from tourist trips to Turkey and Egypt produced by families, couples and singles (for details see Haldrup and Larsen 2009: 54–6). By accounting for how different modes of belonging are 'routed' by particular performances and experiences of distance and connection to/from cultures, people and places of their 'tourist life' I show that tourism indeed makes way for new ways of 'banal cosmopolitanism' but that such cosmopolitan orientations in tourist and everyday life performances indeed at the same time may re-emphasize and (re)produce orientalist scripts. In what follows I want to show how everyday 'ethnographies' in people's private homes shows may contribute to trace out the different modes of 'banal cosmopolitanism' enacted in relation to the afterlife of tourist performances and experiences.

Aesthetic Cosmopolitanism

As Nava (2002) has shown there has been close ties between the cosmopolitan orientations in everyday life and modern consumption. Others have similarly argued that the emergence of contemporary tourism has reinforced a sense of aesthetic cosmopolitanism among certain groups of people (Lash and Urry 1995: 256–7). Thus, in their research on 'cosmopolitan cultures', Szeszynski, B. and Urry, J. (2006) suggests a set of cosmopolitan practices and dispositions consisting of extensive mobilities, curiosity about places and ability to consume and compare places and an openness towards 'other' people and cultures; practices and dispositions that enable, at least certain groups of people to 'inhabit the world from afar'.

Such relations between cosmopolitanism, consumption and tourist travel were also clearly exhibited in the homes of people visited as part of this study. This was perhaps most clearly visible in the home of Keith and Lisa, a young couple living in a small flat in a 'multicultural' inner city neighbourhood in Copenhagen. To Keith and Lisa travelling is a formative part of the way they present themselves. On their shelves are several guidebooks, a globe, a water-pipe, on the wall a map of the world beside a pop art reproduction of a Coca-Cola bottle and on the floor several cushions from Turkey and Latin America. Lisa explains that she likes having cushions from Turkey in her lounge as they are not 'typical, not oriental in their style, you cannot really see where they are from, only I know it, so if anyone asks I will of course tell'. The objects displayed in the sofa on the shelves and so on are not regarded as souvenirs but rather as utility items imbued with a certain symbolic value that enables them to signal Keith's and Lisa's community with an cosmopolitan 'global culture'. However, the personal relationship and knowledge about the origin places of the objects are important to this couple, and travelling together is, they explain, central to them. This dominantly aesthetic ability and sense of being part of a 'global culture' is expressed clearly by several informants in our study so far. Travelling and being part of 'global culture' is important for being able to navigate in the contemporary world. It is also a central competence and source of identity work. To Keith and Lisa the experience of travelling together was central to their trip to Turkey. This was their first time travelling together and this first experience of being away together was marked by literally inscribing

Figure 3.1   Keith and Lisa's lounge 1

Figure 3.2   Keith and Lisa's lounge 2

Figure 3.3   Guidebook with travel routes from their holiday

Figure 3.4   Ann and Pennie's lounge

Figure 3.5   Photo album with pictures from diving in the Red Sea

their routes (and relationship) on the map in the guidebook, they brought with
them (see Figures 3.1, 3.2 and 3.3).

The role of travelling together is part of marking important steps in one's
life, and as part of education oneself and one's children to be part of 'the
global'. Ann had recently finished her Master's degree, the same summer as
her oldest daughter was about to start school. Having travelled extensively in
Europe, Egypt, Jordan and Israel earlier on, Ann explains that she thought this
to be a fine opportunity to let her daughter Pennie gain first-hand knowledge
'other' places: 'The idea was something like: I had been terrible busy writing
my master report ... and then it this seemed like an idea that could be a way
of getting in tune with each other's lives again...before she would start school
... which is ... something big, right...'. The choice of marking the step for the
daughter from child to school child by travelling was not accidental. Ann and
Pennie's home is, like Keith and Lisa's filled with references to other places,
a world map, a canoe, in an extra room, eighteenth century travel literature,
and the wall is dominated by a huge world map, that Ann explains is used for
plotting travels of family friends and 'of course to see Africa, with Egypt as
the reference point' (Figure 3.4). The decision to pick Egypt as the place to go,

however rested on an extensive comparison of the particular affordances this destination offered, Ann continues,

> Well, I thought much about Pennie and Egypt … the desert, camels, that kind of things …. The reason for that I supported the idea was, that I thought…this would be something that would be so much different from … […]. When we were talking about where to go, Paris was mentioned too, but I had it like that, well, no, we can do that another time, when she is a teenager, then she can bring a friend, and let me alone. Because … I simply do not think Paris is able at … to children I mean … leave an impression … Whereas I thought, Mummies are potent, the big temples, the desert, camels and the smells, and like this, there are many things inviting that you can construct … mark that this is an extraordinary experience you have had …

Despite the many references to 'other places' in her 'lounge', no souvenirs are on display (apart from the statue from the Dali Museum in Figueres, Spain in the corner). When asked about this, Ann explains that she earlier on bought loads of souvenirs but that she often feels a bit ashamed of the things she has brought home (several jars (replicas of jars used by ancient Egyptians for mummification) a chessboard, a backgammon board and small statuettes of Egyptian gods).

> … I think I brought them home to document the history of Egypt, statues of Akn-aton, papyrus that sort of things. I did think they were fantastic from an aesthetic point of view, but also to show them at home, a kind of documentary: See, that's Egypt …. I wished to document it to my family, especially my grandparents who are very local … then I could like go out, experience, document and tell them about…

Translating her travel experience into the local life worlds of – in particular – her grandparents was an important enterprise earlier on. Today it is the transfer of the competences and identity of 'being able' to get along while travelling that is central. Asked about how she holds on to her travel experience of this trip, she emphasizes – again – the educating elements of travelling to her daughter. Thus, Ann's, primary contribution to holding on to their collective travel experience is producing a photo album (planned from the very beginning) she edited and finished for her daughter to her seventh birthday. The scenes in the album are portraits of Pennie in different situations: Pennie, leading the mountain guide in Sinai, Pennie at work in the restaurant with the waiter, Pennie above the 'blue hole' coral reef in the Red Sea – scenes that Ann explains she refers to, when something puts her daughter down (Figure 3.5). The deliberate staging, choreographing and photographing of a 'happy childhood' while on holiday is of course a main aspect of all family life (see Haldrup and Larsen 2003, Larsen 2005). What interests me here is how this social endeavour is related to the 'cosmopolitan outlook' emphasized by Ann and the socialization of her daughter to include cosmopolitan connections in this photo narrative. Openness towards 'other people' and their lives (children in the Bedouin camp, where they stayed a couple of nights, waiters at the café close to their hostel and so on) and the ability to map her own 'place' and culture onto the globe are made central elements of the daughter's identity and education.

Orientalist Cosmopolitanism

As more and more groups in society have gained access to cheap air travel around the globe the significance of the ability to experience, to compare and discriminate between places and cultures are not reserved for the young and educated. Such cosmopolitan values were also a central theme for Heinz, 60-year-old, retired and living in a small, remote village. Since his first tour outside Europe in 1994, just after the younger son moved, he has been a frequent traveller. Before my first visit to his house in November 2006 he had already been to the Philippines, Egypt and Bali this year and was just packing to go to Thailand the day after my visit. Heinz keeps an archive of systematic reports of the places he visits. Every evening on his travels he fills one sheet of paper with his impressions from that day, the sheets are ordered into the same blue folders, and kept in the cupboard besides the television together with tickets, photocopies of travel brochures from tour operators and so on. He also documents his holiday on video tape. He explains this systematic recording of his travels as partly 'because I like to compare what they write and how things are' partly because 'it is becoming difficult to distinguish between, different sorts of things, remember how things were in a specific country in say 2004'. After some talk Heinz explains that the reason for the documentation originally was (and in some sense still is) that he would like to be asked about his rich travel experience especially by his sons, but also from friends. In emphasizing the importance of being a skilled and experienced traveller (he also organizes most of his travels himself), and the seriousness that goes into comparing and evaluating places he visits, Heinz explicitly emphasizes the value of cosmopolitan orientations in everyday life. The ability to organize travel, distinguish and evaluate places, cultures and people as well as to appreciate difference in global society is all high ranking skills and values. However it also becomes clear from the talks with him that not all places, cultures and people are equally compared. For Heinz places are assigned very specific roles, relating to their particular affordances. This is also mirrored in the repetitive patterns of his travel career: Thailand is for relaxing, Bali for exoticism and Egypt for diving and so on (see Figures 3.6 and 3.7), and when speaking with Heinz it also becomes clear that his cosmopolitan outlook goes hand in hand with a geographical imagination based on orientalist tropes. As Gregory (2005) has argued the theatrical qualities of the Orient have been (and still are) central in orchestrating how tourists have encountered, perceived and portrayed other places, cultures and people also when performing tourism, and also in connecting with places, cultures and people

1994 Tunesien
1995 Santorini
1995 Kos
1996 Marganita
1996 Autocamper
1996 Teneriffe
1995 Florida
1997 Malaysia
1997 Typhiset
1998 Thailand
1998 Rhodos
1999 Thailand
1999 Bali
2000 Egypten
2000 Filippinerne
2000 Bali
2001 Mexico
2001 Egypten
2002 Gili Trawangan
2002 Egypten
2003 Filippinerne
2003 Brasilien
2004 Thailand

2004 Bali
2005 Filippinerne
2005 Egypten
2005 Bali
2006 Filippinerne
2006 Egypten
2006 Bali

Figure 3.6   Heinz's travel career

Figure 3.7   Travel itineraries and artefacts

'outside' the geographical East (Sheller 2003). As argued elsewhere orientalism is not only a regime of (representational) knowledge but also an embodied, tacit and habitual knowledge that shapes people's encounter with 'O/other's by translating hegemonic discourses into everyday practices so that they enter into the habitual spaces of ordinary experience. It is about the way in which cultural difference is performed in the encompassing field of everyday sociality and sensual habit – how it colours the visual, flavours the olfactory and tempers the emotional (Haldrup et al. 2006). The roles Heinz assigns to places are firmly rooted in such an embodied, sensual and habitual practical orientalism. This is most clearly in his distinction between Egypt (where he only goes for diving trips) and Bali ('an inviting island in a sea of Muslims'). He talks passionately about his long relationship with a Balinese family and especially their daughter, 17, at the same time explaining 'no, I do not talk to people in Egypt. You know I have only met the diving instructors and they are young men, and our everyday lives would be too remote from each other to have anything to speak of …'. Heinz does not like speaking about this topic, when asked directly, but he will touch upon it occasionally when we are talking about other issues. For example, when discussing the cartoon controversy discussed in the introduction of this chapter, he argues, 'well, I do not want to accuse all Muslims, the diver instructors for example are absolutely fine, but …'. Or speaking about debates on multiculturalism in Denmark 'well, I do not want to project the debate we have in Denmark onto the people living there, but it may be something that sticks to you …'. In this way, we see his apparent cosmopolitan orientation is ironical, informed by a differentiation between and evaluating of places based on orientalist tropes.

The workings of such an *orientalist cosmopolitanism* is even more significant in relation to another couple: Dennis and Maryann. Speaking about their trip to Egypt Dennis explained that they had to stay in the resort area or on their cruise ship because when leaving these patrolled spaces 'I always made sure to be together with some other people. (…) I would say it is really frightening. You cannot move anywhere in Egypt without passing road barriers and barb wire …'. To Dennis and Maryann the streets of Egypt are terrifying with potential assailants in every street. The main model of explanation used by Dennis is not cultural but economic. He bases this interpretation mainly on talks he has had with a member of the crew on his Nile Cruise ship ('his friend') also using this as a way of emphasizing to me that he is not prejudiced against the people he meets. Nevertheless he generalizes the 'poverty' explanation to a general model of interpretation both used in relation to his visits in both Turkey and Egypt. Thus, Dennis continually referred to the greedy and swindling nature of people in Egypt and Turkey and while these are of course classical tropes of orientalist discourse (Said 1995) he also relates this to the experience of living in a multiethnic society at home: '…but a big difference between being in Turkey and Egypt. In Turkey, the people who have been living up here for some time, they are really disgusting, when they are in their homeland. Unpleasant like hell …' and asked examples of incidents of this he explains '… its incidents like, if you do not want to look at his goods, then you are met with … and shouting hooker after your wife and stupid pig … and other abuses'.

To Dennis and Maryann the world is one big buffet. They are in many ways emblematic of Bauman's notion of the world as 'the tourist's oyster' (1998). However, their trips to the different corners of the world, rely on a sharp dichotomized division between safe/unsafe spaces, 'them' and 'us'. Streets are filled with unreliable, lurking, potentially violent assailants and safe only in the bus or accompanied by 'other people'. Another woman in her early thirties similarly describes how she and her mother were constantly yelled at by shop owners shouting 'there comes the whores again' and assaulting her by putting ice cubes inside her blouse and bra. To her, the streets were disorienting spaces, and the behaviour of the 'local' people inhabiting them was confusing, assaulting, and basically impossible to relate to. Like Dennis, she relates these incidents to her own experiences of multiethnicity at home, claiming that '… but I knew this, I know how it is in my youth club (her work), when there is an arrangement of some sort, then they turn up in groups from Gjellerup Parken [a renowned Ghetto area in the city she lived in]. And they do not have any respect for anything … especially not the way they approach women …'. Through their embodied and sensual coding (sound, smell, mere presence of 'alien' bodies etc.) everyday spaces become battle zones between hegemonic western bodies and perceived threatening, anonymous 'aliens'. Furthermore, this geographical imagination is translated into the streets of the bazaar producing a space war, in which the tourists only are capable of moving around like an occupant power patrolling the hostile streets in groups and with guards capable of making immediate retreats to the 'heavens' provided by the hotel, the cruise ship and so on. Like this we see that explicit cosmopolitanism is not necessarily incompatible with practising orientalism in the ordinary encounters with places and people on holiday.

Connective Cosmopolitanism

As we have seen above, aesthetic appreciation and practical distancing go easily hand in hand when doing tourism. In doing this, tourist performances and experiences tend to reinforce a sense of 'orientalist hegemony' of the Western, rational, mundane subject over the Eastern, emotional, exotic object. It should, however, also be noted that tourist experiences also may challenge and alter such orientalist schematics. Reflecting on the relation between tourist experiences in Turkey and Egypt and everyday encounters with immigrants from these particular countries some of the people visited told that their experiences on holiday had transformed their view on immigrants from what we have called a habitual embodied 'practical orientalism' to a more open attitude. This transformation was made particular clear by Ken and Sue. Both had lived in so-called 'ghetto areas', stigmatized for their high proportion of Middle Eastern and Third World immigrants, and worked as factory employees together with people of multiple ethnic origins. When talking about how tourist experiences on their frequent holidays in Turkey related to broader aspects of their everyday lives, Sue emphasized this:

> But don't you remember in the beginning … the first time we were going to Turkey … we were walking through the shopping centre and there … with all these Turks hanging around everywhere … and you said: "what the hell are we going to Turkey for. We've got plenty of these people here '", and Ken continued: 'Yeah, "Well, it is all those people they don't want to keep down there that they are sending to us up here …" […] And then when you actually get down there it is plainly speaking like getting a basket of coal thrown in your face … Well what is it actually you've been saying …

After having mentioned some of the instances in which their everyday conceptions of Muslim immigrants has been changed by their experiences as tourists in Turkey. Ken sums up:

> I would not say that I look differently on Turks here. I think I'll say that I might view them more like I do when I am holiday … It is not like it was before you actually got to know someone from Turkey … […] There is of course some rules you adjust to when you are out there…Actually you do not really think about what you bring home, but if you bring these with you home […], then it actually works out just fine….

To Ken and Sue the holidays in Turkey have, in important ways, changed their way of relating to and encountering people of 'other' cultural backgrounds. Thus the practical orientalism which has shaped their relations to immigrants earlier in their life has been superseded by what could be called a connective cosmopolitanism based and situated on experiences and practices of interaction with particular 'others'. The notion of connected cosmopolitanism is close to Appiah's suggestion of '… a world in which everyone is a rooted cosmopolitan, attached to a home of his or her own, with its own particularities, but taking pleasure from the presence of other, different, places, that are home to other, different, people' (Appiah 1998: 91). Like Appiah's notion of 'rooted' cosmopolitanism the notion of 'connective cosmopolitanism' forces us to think of cosmopolitanism as something that relies on and reproduces a sustained commitment to particular places and people instead of taking pleasure from being part of an abstracted global or cosmopolitan culture.

This also differentiates this specific mode of cosmopolitanism both from the positive valuation of 'otherness' in aesthetic cosmopolitanism and the highly negative valuation of the 'other' found in the culture of what I have called orientalist cosmopolitanism. While the latter two both draw respectively on a pleasurable or repulsive distance between well-known/exotic, tourism/everyday life, tourists/ locals, Self/others connective cosmopolitanism is based on specific encounters and abilities to connect with other particular people, places and cultures. This can be further elaborated by turning to how tourist experiences and everyday life are tied together in the life of Hanna.

Like Ken and Sue, Hanna does not have a very long travel career. She has only been to the Canary Islands with her husband a couple of times and a visit to some family in Florida back in 1991. The visits to Egypt all have been connected to her and her husband's purchase of a house in Hurghada on the Egyptian Red Sea shore. As her husband has been too busy with their business in Denmark, she has had to supervise the constructing work in Hurghada. At the same time her visits have facilitated friendships with people there and in Cairo, friends that have been inviting her occasionally to family events and on visits to their own families in suburban Luxor, Cairo and Alexandria. Speaking about how she has got in contact with the people that populate her photo album she explains 'Well, they are very eager to seek contact, you know, on the street, so some of them we just met, and if I then feel that they see alright, I do not have any second thoughts about letting them suggest their ideas …[…]. And I am not worried about … that, you know they like to invite you to your home, and show what they've got … so it's simply to follow … […]'. When asked about what kind of conversations she engage in and what topics she brings up she explains, that she does not have any problems regarding political discussions:

> We do have sustained contact … We should actually have been to Egypt in February when all flights suddenly were cancelled [as a consequence of the 'cartoon crisis']. And how I felt about it …. Well, I was very curious about what they thought about it all, in Cairo. We were in daily contact … and I was thinking about what he was thinking about it, so I did send him a text message, just to hear what the guy we have most contact to thought about … and I got a very fast reply saying that we are fully aware of that not all Danes are like that … so no, it is very reasonable … of course some have been critical … but we were there just afterwards, but no one commented on it to us … When we asked … they were fully up to date about the discussion, but no ….

She has regular contact with her 'friends' (two in Cairo, two in Hurghada) at least every fourteen days, mostly about everyday issues such as work, and progress with the house, but also at moments such as national holidays (where her Egyptian friends are very eager to remember the central holidays such as Christmas and Easter), or when incidents such as the cartoon crisis or bombings (Sharm, Dahab) hit the global media circuit (Figures 3.8 to 3.10). Apart from that, communication revolves around trivial issues such as their respective children, how business is going and so on. Asked if there is anything she misses in her relations to the 'Egyptian' friends, she says:

> I miss talking more about the role of women, you know, the women I know, do not really speak English, so their men have to be there, and I know that for example … Musa's wife is fed up, with him claiming that she should stay at

Figure 3.8   Home of Hanna and her family in Northern Jutland, Denmark

Figure 3.9   Second home under construction, Hurghada, Egypt

Figure 3.10 Family visit to Egyptian friends, Cairo, Egypt

Figure 3.11 Susan's computer, webcam and photo diary

Figure 3.12   Family and souvenirs on display in Susan's home

home, and I try to explain that he should let her have some of the freedom he
also wants for himself, but it is difficult....

As we see the 'cosmopolitan' orientation of Hanna is not based on aesthetic
premises but much more on the practicalities of everyday life. While being
'tourist by accident' (the house in Egypt was originally planned as an investment
opportunity), Hanna was not particular interested in 'experiencing' spectacular
cultures and sights, the connections she keeps alive are very much based on
comparing and discussing the differences and sameness of their everyday lives.
Instead she emphasizes the ability to connect and make friendship across cultural
and geographical distance and to reproduce such connections in everyday life
communications between Denmark and Egypt.

A similar connective cosmopolitanism is also present in the life of Susan.
Despite travelling to Egypt regularly as a two-week sun-seeker, Susan explains
that she prefers to avoid any contact with other tourists and travellers and even
claims that she has never gone on a sightseeing trip. 'I don't care ... simply too
sick ... just about pulling the last money out of the tourists ... then I will prefer
to sit at the local coffee shop, have a pipe of tobacco, and a cup of Arabian coffee
... I mean ... that's much more cosy ...'. Instead she refers to particular people
in Egypt that she has a continual contact with also in-between her visits. Asked

about what topic she brings up she explains that it is the same type of discussions she would also engage in at home 'for example divorces ... when I discuss with other women ... [...]. Now take these cartoons as an example, they tend to laugh a bit, but I guess that they are not the worst kind of ... people working in the resort areas...they are much more engaged in supporting their families... So we discuss divorces, politics and so on ...'. During her visits she (and her former husband) have been involved in different projects with local entrepreneurs such as the design of a polygon ice house in Sharm el Sheik during their first visit.

Over time she has developed relationships with a couple of Egyptians working in Hurghada, one of them having been in Denmark too to visit his family and she has continual contact with these using Skype and text messages on the mobile phone. In doing this they are included into her daily communication practices (see Figures 3.11 and 3.12). With close relatives living in seven different countries on three continents, using Skype is an integral part of her social life.'Uncle and Aunt in Haifa, my other uncle in Canada, and my father has just been married and moved to Bucharest'.

Her home is filled with souvenirs bought on her visits, including several places between pictures of family members and so on, and she tells them that using these objects – such as the oil lamp from Israel – makes her think of her family relations around the world. While not being especially affluent – she is a lone mother, currently working renting out skates at winter, running an ice cream bar in the local park in the summer – her small apartment is stuffed with objects and pictures pointing to the 'connective cosmopolitanism' which orchestrates her life.

Conclusion: Tourism and Banal Cosmopolitanism

In this chapter I have shown how tourist performances in significant ways feed into the practices and relations of peoples' everyday lives, connecting spaces of tourism and spaces of home through a variety of corporeal, virtual and material mobilities. Tourist performances are not limited to the two-week vacation of the pleasure-seeking traveller, but integrated into wider aspects of consumption and everyday life inducing processes of 'banal cosmopolitanism' into peoples' lives. As we see from the examples discussed here, the particular routes through which cosmopolitan experiences may travel are not univocal. Thus, cosmopolitanism may take different, even antagonistic forms. The practices and dispositions developed and employed by the households in 'routing' their cosmopolitanized practises of 'away' and 'home' in their tourist performances are very different. In addition to the 'aesthetic cosmopolitanism' which often has been noticed to be an integral part of the rise of modern tourism, I suggest the notions of 'orientalist cosmoplitanism' and 'connective cosmopolitanism' to capture the different modes of banal cosmopolitanism at work in a world in which the 'global' is increasingly part of mundane experiences. Thus, the notions of 'orientalist cosmopolitanism' and 'connective cosmopolitanism', both point to more non-reflexive and embodied responses to the penetration of everyday life by global mobilities than what is usually addressed in discussions of cosmopolitan cultures. By discriminating between such modes of cosmopolitanism I want to emphasize that apart from an emerging culture of aesthetic cosmopolitanism among particular asocial groups, other modes of cosmopolitanism, perhaps less reflexive and less articulated, may be enabled by current transformations of everyday life and travel. Thus processes of banal cosmopolitanism may produce borders and connections, that are drawn and established, in habitual, non-reflexive and embodied ways.

Distinguishing between such differences in the way people respond to the presence of 'the global' in tourist experiences and performances is furthermore important if we want to avoid the hyperbolic banalities that often haunts discussions of globalization, tourism and cosmopolitanism. Tourism does indeed carry emancipatory potential as I quoted Löfgren for stating in the beginning of this article. The way this potential is embodied, particularized and materialized is however an open game.

References

Appadurai, A. 1999. *Modernity at Large. Cultural Dimensions of Globalization*, Minneapolis: University of Minnesota Press.

Appiah, K.A. 1998. Cosmopolitan patriots, in *Cosmopolis. Thinking and Feeling Beyond the Nation*, edited by P. Cheah and B. Robbins. Minneapolis: University of Minnesota Press, 91–116.

Bauman, Z. 1998. *Globalization. The Human Consequences*. New York: Columbia University Press.

Beck, U. 2004. Cosmopolitical realism: on the distinction between cosmopolitanism in philosophy and the social sciences. *Global Networks*, 4(2), 131–56.

Billig, M. 1995. *Banal Nationalism*. London: Sage.

Clifford, J. 1997. *Routes: Travel and Translation in the Late Twentieth Century*. Cambridge, MA: Harvard University Press.

Gregory, D. 2005. Performing Cairo: Orientalism and the city of the Arabian nights, in *Making Cairo Medieval*, edited by N. Alsayyad, I. Bierman and N. Rabat. Lanham, MD: Lexington Books. 69–93.

Haldrup, M. and Larsen, J. 2003. The family gaze. *Tourist Studies*, 3(1), 23–46.

Haldrup, M. and Larsen, J. 2009, *Tourism, Performance and the Everyday*, London: Routledge

Haldrup, M. and Larsen, J. 2006. Material cultures of tourism. *Leisure Studies*, 25(3), 275–89.

Haldrup, M. Koefoed, L. and Simonsen, K. 2006. Practical orientalism – bodies, everyday life and the construction of otherness. *Geografiska Annaler B*, 88(2), 173–84.

Larsen, J. 2005. Families seen sightseeing: performativity of tourist photography. *Space and Culture*, 8(4), 416–34.

Löfgren, O. 1999, *On Holiday: A History of Vacationing*. Berkeley: University of California Press.

Nava, M. 2002. Cosmopolitan modernity: everyday imaginaries and the register of difference. *Theory, Culture & Society*, 19(1–2), 81–99.

Said, E., 1995. *Orientalism*. London: Penguin.

Sheller, M. 2003. *Consuming the Caribbean*. London: Routledge.

Szeszynski, B. and Urry, J. 2006. Visuality, mobility and the cosmopolitan: inhabiting the world from afar. *The British Journal of Sociology*, 57(1), 113–31.

Turner, J. and Ash, L. 1975. *The Golden Hordes*. London: Constable.

Chapter 4

The Island That Was Not There: Producing Corelli's Island, Staging Kefalonia

Mike Crang and Penny Travlou

Introduction

This chapter will focus upon the contested practices and imaginations of one island whose tourist market is markedly divided between an upmarket north and mass market south. In the midst of this tense clash of tastes, the island was the setting for the book and the film of *Captain Corelli's Mandolin*. So this chapter moves between the Louis de Bernières' book *Captain Corelli's Mandolin* (1997), the Miramax film of the book (released 2001) and the touristic experience of the island. In the year after the release of the film visitor numbers from the UK to the island, who form some 87 percent of those arriving by plane, rose by 22 percent and 10 percent again the year following, more strongly than growth in visitors from other countries, and growing more rapidly than British tourism to Greece in general (Hudson and Ritchie 2006: 263–4). It was by all accounts a classic case of a movie driving up the popularity of a destination. This has set in play competing and complementary imaginaries of the island as landscape and beach resort – and what such beaches should be used for. Hosting the (so-called) most photographed beach in Greece, alongside the beaches, or 'coves', labelled as 'romantic' via the movie, alongside mass tourism infrastructure the chapter unpacks the production of the beach and scenery for tourists. Not least here we want to highlight the different Kefalonias imagined and those lost and found, those unobtainable and those haunting the Ionian.

An Island Lost in the Myths of Time

'On an island untouched by time' intones the gravely baritone voice in the first words of the trailer for *Captain Corelli's Mandolin*, accompanied by brief scenes of beaches and fishing coves and barefooted women on the sand and then followed by the speeding sequence of war planes flying over and troops marching. 'In a world divided by war', it continues, as Italian troops march past Captain Corelli (Nicholas Cage) who, leading his unit, calls out 'Bella Bambina 2 o'clock' and leads the salute to Pelagia (Penelope Cruz). 'A stranger came into their lives' adds the narration. We shall not tarry over each scene, but here we find the film quickly setting the island of Kefalonia as a pristine, edenic land – out of time, out of the political world into which events roughly intrude. This is clearly a replayed scenario of so many island fantasies. But it is also one that resonates most strongly with the Greek Ionian pitched in terms of Otherness to modern Europe.

As Tzanelli notes 'Greece has been more an imagined *topos* than an actual place for prosperous foreigners since the birth of classicism' (Tzanelli 2003b: 21). Originally this focused upon the role of ancient Greece as progenitor and originary mythic place for European, Western civilization. But in those visits and studies increasing concern was then paid to the neo-Hellenes, the modern Greeks – portrayed by contrast as oriental interlopers; so the imagined superiority of ancient western Greece could be revealed in the economic inferiority of the contemporary Greek peasantry. Indeed, in the case of Van Lennep's *Oriental Album* – writing in the 1860s – Greek culture is assigned to Ottoman and Oriental categories: 'it revolves around the typical oriental themes of exhibition and colourfulness' (Tzanelli 2003b: 22). However, there certainly was a more celebratory if patronizing subtone focused upon the rural peasantry, celebrated for their earthy if simplistic culture – seen as pure and authentic as opposed to hybridized. Accounts reveal partly an inevitable fascination with the inaccessibility of Greek women for observation and focus upon 'gestures, posture and mannerism – *neo-Hellenic habitus*' (Tzanelli 2003b: 40, 2003a). This is part of a piece with a rising romantic celebration of timeless folk culture, clashing with the linear evolutionary temporality seen in developmental models. Such temporalities map onto an urban rural divide:

> Urban time emulated western progressive time unsuccessfully, whereas the rural was almost static. Greek rural culture was burdened by a famous past, which was preserved *unconsciously* in popular beliefs. Contrariwise, there was a violent historical disruption in urban modes of thought. (Tzanelli 2003b: 43)

If this trope is clearly being tapped into by the movie, it contrasts with the novel whose main strand includes precisely a layered historical sensibility embodied in the figure of Dr Ioannis and his incomplete manuscript for a 'New History of Cephallonia'. Through this textual vehicle, de Bernières is able to provide a meta-history of the island in the novel. Here is a history that is anything but uneventful, and is rather written as though the island were endlessly buffeted by the winds of geopolitical conflict – from ancient times, through the Venetian empire to Ottoman domination and British imperialism. It provides a cataclysmic history of disasters and explosions, of opposite outcomes of good intentions, sceptical of metanarratives (Sheppard 2002). But within it there is also a clear sense of a replication of the stereotypes of Ionian culture produced through the British colonial moment and its ethnologies – offering a romantic folk type.

Yet the novel can clearly play more upon this setting, where it ironicizes other Ionian myths. In the *Odyssey*, faithful Penelope stitches at home by day and undoes her work at night. Pelagia ends up unstitching her work and growing cooler to her suitor whilst Mandras, away fighting a war, returns and is only recognized by the pine marten (similar to the dog who recognizes Odysseus in the Odyssey). But then Odysseus leaves his sailor with the rotting leg whereas Mandras returns with a rotting foot (Sheppard 2002). In other words, on a deep level the novel inserts the island in an economy of long-circulating stories and images of the island.

As such we might look at the long term envisioning of the Adriatic in things like mid-twentieth century travellers' home movies where in the interwar years 'cinematography – as enthusiasts called their new hobby – was part of the new technological apparatus that, together with locomotives, cruise ships and motor cars, became associated with Mediterranean holidays' (Norris Nicholson 2006: 16) and indeed the staging of a folksy Greek persona in the post war boom in musical dramas (Papadimitrou 2000). The Greek island thus has a long popular and academic history as a space of constructed visibility. Thus as Tzanelli (2003) points out the book and film both refigure the island but also incite a desire to visit or explore that island, they are doing so as part of the ongoing discourse of Greekness, Mediterranean-ness and indeed the nature of holidays. One important reflection here then is going to be between book and site, between imagination and representation as we see these imaginaries play out. There has been a small boom in work examining the intersection of film depictions and locations and tourism. Much of it started from a concern to measure the possible impacts of films upon visitor numbers (e.g. Tooke and Baker 1996; Riley, Baker et al. 1998) and indeed there has developed a small industry of conventions and promoters seeking to attract films to use locations as a place marketing tool (Hudson and Ritchie 2006). The media effect is not restricted to cinematic films (witness for instance the reported boom in 'Toddler-tourism' to the village of Tobermory on the island of Mull after it was used for the setting of the popular pre-school show *Balamory* (Connell 2005)). Rapidly, though, two sets of issues emerge in the literature. Perhaps first, is that of the malleability of location, second the multiplicity of motivations.

On the first, it is very quickly apparent that destinations linked to films or media can be plural. We have cases where the original book or similar may use a real setting, and then the film chooses to use a different location to represent that. Thus in the Mediterranean, the film *Troy* has been credited with a boom in visitors to Malta – where the movie was shot. Of course the original setting may be fictional and the film has then to use a real setting to stand for that – perhaps most famously with the *Lord of the Rings* trilogy, where Aoteoroa came to stand for the imagined Middle Earth itself fused from the refracted Celtic imaginary of English myth. Yet that is also far too simple, for there may well multiple original locations (say, for Jane Austen's settings which were amalgams inspired by several country houses) and these in turn may well be represented by fragmented settings (where one location is used for interiors, another for exteriors) in ways that can jump-cut time and space (Crang 2003) where, for instance, Harry Potter tourism extends to an arc of 300 miles to cover the key places marketing themselves as locations in the film, or where tourist trails for 'Inspector Morse' deliberately stage some of the impossible entrances and exits that teleport the detective across town, entering one site to be seen emerging from another. Meanwhile more evergreen stories may appear in multiple media using multiple locations over time. Moreover, multiple different stories can blur through the use of the same area. And last but not least, it is entirely possible for tourists to mistake the location entirely and visit somewhere else convinced it is related to a film or story. The point is in some ways banal, though in terms of the layered and contestable meanings less so, but it is important. One reason for the focus of this study was that book and film location coincided on the island of Kefalonia – though up to the end, so the local story goes, the film was nearly shot on Corfu.

The second complicating factor tends to be taken as the extent of motivation that the film supplies for different tourists. Thus a number of writers have quickly divided up movie tourists into those we might call 'happenstance' – they happen to be there and thus show an interest, through to the 'generally aware' who show a specific interest, thence to the more and more committed who travel specifically motivated to see things about that film overlaid on which are the different motivations afforded by the film be they plot related, personality related (the stars, and so on) or place related (the scenery) (Macionis 2004). Quite crucially for our story here, the reviews of *Captain Corelli* tended to respond 'the Island is the Star', and one suggested it turned 'Cephalonian nature into the principal actor of the story' (Hudson and Ritchie 2006: 265) or rather more acidly one suggested that the landscape was the second great actor after Cruz's eyes, and 'all those sun-drenched shots of white beaches and "colourful" villagers will do wonders for the Cephalonian tourist industry' (*The Daily Information* 4 April 2001). Clearly movies where the place features more strongly should exert a stronger pull to tourists. The academic studies then tend to work to develop typologies and scales of motivations in more refined details – connecting typologies of visitor motivation between films and literature and so forth (Busby and Klug 2002). However, only rarely do they consider the affective qualities of the image (Kim and Richardson 2003) – it tends to be treated simply as product placement, where merely seeing the destination is enough, rather than thinking through the registers of affection or desire through which it is attended.

Meanwhile on the other side are a range of social theoretical statements that seek to connect to theories have argued that tourism necessarily 'spectacularizes' destinations or creates 'myths' about them. John Urry's (1990) Foucauldian notion of the tourist gaze, that shapes perceptions of the landscape, has been enormously influential. Put over-simply, critical approaches suggest a vicious hermaneutic circle where marketing or films create a desirable image, which tourists then reproduce in their own pictures and memories (Crang 1997). Thus people do not see the landscape but rather see it as filtered through media-led expectations they then confirm. These all highlight how media inscribe meaning onto places with representations and texts acting as signifiers and markers of things which tourists should travel to see. The double edged issue is that media are seen not only investing places with meaning but also limiting them to a script, reducing them to simplified images – that can be captured and reproduced – where culture is formatted for easy transport as 6 inch by 8 inch pieces of glossy paper. The result is a reduced but parallel vision of landscape through both media and tourism (Acland 1998). Thus the indexing of places to cinematic images often leads to disappointment or anti-climax as sites do not have single or original meanings and moreover tends to homogenize the viewed 'other' and only allow heterogeneity to us – they are unproblematically represented as mirrored on screen (Galani-Moutafi 2000: 213, 219). There is then in this critical view a connection between hegemonic ways of seeing the world encoded in both visual media (especially film and television) and tourism as being increasingly intertwined forms of spatial appropriation (Jansson 2002), and these ways of seeing are often suggested to distance people from the world rather than engage them via other affective registers. In Jansson's terms they are antagonistic modes of appropriation, that set us apart from the landscape viewed as part of a 'tourism phantasmagoria' generated through tourism imagery. Jansson suggests disentangling more sympathetic modes of appropriation and indeed context appropriations (focused on activities where the place is more incidental). But in this paper we focus rather more upon the phantasm – asking about the substantiality of Corelli and the imagery itself.

Anticipating Captain Corelli's Island

An analysis of UK-based brochures for the Mediterranean shows pictures dominated by blues and whites, and in the Greek sections and those on the Ionian islands, under the headings 'ideal for couples', romantic destinations and family destinations, the introductory pieces on Kefalonia mobilize a common set of tropes:

Castaway Kefalonia - the island of Captain Corelli fame (Thomson 2005)

As fans of *Captain Corelli's Mandolin* will undoubtedly know, Kefalonia consists of peaceful bays, tiny hillside villages, sleepy harbours and, also, some wonderful beach resorts (MyTravel 2005).

Kefalonia is a haven for beach lovers with its sand and shingle coves, sheltered bays and inlets. … still relatively new to mass tourism, although it has become famous due to the success of the book and film 'Captain Corelli's Mandolin' … Sami 'if you want a quick preview see the movie 'Captain Corelli's Mandolin' that was filmed here (Thomas Cook 2005).

The setting for the romantic story of Captain Corelli's Mandolin, this mountainous isle is the largest of the Ionian cluster. Cliffs and caves, picturesque little ports, sleepy villages in herb scented hills, and beautiful beaches – some with watersports, all combine to create the perfect place and space to chill, unwind or enjoy a family holiday. The old Greece with modern comforts (Airtours 2006).

Of course any study of brochures has to set that in the context of the dominance of the pages by pictures and details of pretty standardized accommodation (Dann 1996), and here each 'destination' is an island and within it a number of resorts mean that you have a pattern of 1 page setting the scene for the island then 4–5 pages of accommodation. So the sense of the destination here blurs from Greece to an Island, to a specific resort. In that context Kefalonia however is notable for the scenic and landscape descriptions that are often entirely absent from other destinations. The island truly is the star. And it is the star in the living room before arrival – it is even, as Thomas Cook say, possible to preview your trip via the movie. When the movie was released Greek specialist tour operator Kosmar and East Midland Airport took 90 Midland travel agents to a special showing of the movie in order to promote the destination (Hudson and Ritchie 2006: 263).

The island of the movie thus wraps over the island of the tourist imagery. But the touristic imagery also already enfolds the movie. Thus the stunning beauty of the island, as mentioned above, is one of the main tropes in otherwise pretty dire reviews for the film. Thus the *New York Times* review (2001) is a virtual paean to the island as a visual spectacle and visual spectacle as part of tourism:

> If you've been longing to visit the Greek islands but haven't the time or money to make the journey, you could do worse than spend a couple of hours soaking up the scenery in "Captain Corelli's Mandolin." Filmed largely on Cephalonia, … the movie shimmers with a bluish-gold luminescence reflected from the turquoise waters of the Ionian Sea. This light lends the craggy landscape a hot coppery radiance that seems to emanate from inside the earth. Cinematographically (John Toll supervised), the movie is a glorious ode to the sun-baked island on which it was filmed.

Here the power of visual imagery is clearly aligned with an assumed connection with tourism. A similar connection is highlighted in *Entertainment Weekly* (2001) whose review opened with the deliberate commentary on the island as tourist destination: 'Sunlight kisses the Greek island of Cephalonia so tenderly in *Captain Corelli's Mandolin*, you'll want to book your next vacation there. In fact, this ad-agency-like adaptation of Louis De Bernières' 1994 book-club favorite turns the historical novel into a travel brochure'. Here then we start to see the folding back of tourist imagery into the movie – now as a critique, that this is too simply replaying clichés of tourist visualization. Thus it points out the positioning of locals as folk colour where 'the Greek townsfolk in this history-inspired story glow with ethnic pride. Then they dance. Life is hard but photogenic', before finally decrying the lack of depth as related to touristic aesthetics 'the passion between Corelli and his Pelagia is indistinguishable from the affection the captain demonstrates for his mandolin. On this island, theirs is a tourist kind of love'. Channel 4's final verdict was 'A fine holiday ad but a rather dull movie'. A similar line, though more trenchantly put, can be found where the world socialist website reviewer Richard Phillips indeed used the scenic qualities to dismiss the movie as 'touristic':

Madden has produced a two-hour picture postcard of Cefalonia's winding hillside tracks, pretty villages, golden beaches and turquoise blue seas. The film, which has apparently boosted tourist visits to the island since its release, provides viewers with little understanding of the period. No effort is made to explore the political and emotional motivations of the film's protagonists.

Here image is seen to be covering historical narrative and contention. What is repeatedly remarked upon as unsettling this visual feast of the Mediterranean pastoral is the sound of it. Not, to be sure, the opera led score, but rather the creation of what Peter Preston in the *Observer* (6 May 2001) called a fused 'zorbaspeak' of 'eastern med esperanto', and Peter Bradshaw for *the Guardian* (4 May 2001) decided made for a light 'holiday romance with silly voices on the beautiful sun-kissed island of Cephallonia'.

Thus we have the damning of the movie for being touristic and the scenery read through that as picture postcard encoded tourist imagery. The film does not escape its island. Likewise the assumptions are that this will automatically play into producing increased visitor numbers. So the question this raises is how the imagery from the film connects with the tourist aesthetic and the practice of the tourist gaze on the island. And to answer that, we wish to turn to perhaps the most iconic site on the island.

The Beach

Kefalonia itself trades upon a beach that has become delocalized by its ubiquity. Regularly appearing as 'the beach' in national campaigns is Myrtos beach. It is delocalized in that it appears unnamed and unspecified, and appears in a different view than it does when it was used for the Italian officers' picnic scene in the movie. The conventional shots taken from above and to the North show a squarish bay between massive cliffs in which circulates eye-wateringly blue sea, stirring plumes of milky white sediment before lapping on a blisteringly white beach. While the image is used nationally and ubiquitously, in terms of visiting the island it is simultaneously and repeatedly inscribed in place, with signposts for car hire outside the airport using the beach as the symbol of the island, and more prosaically road signs greeting travellers with the announcement of the impending approach to 'The famous Greek beach' at a mere 25 km or so distance. Indeed then the road is set up with a special viewing point from which you may view the (famous) view of the (famous) beach – safely 5 km travel from getting your toes wet, and at a platform now rather safer than simply stopping on a blind bend rounding a mountain spur, though many happily strolled across that road each day. And if you hang around that viewing point, as we did, taking hour long samples, in an hour you might expect to see 18 groups of people stop, including two coach parties, but never more than two at a time thanks to the careful scheduling of different companies, for an average of four minutes for independent travellers and a little longer for coaches to allow for disembarking and re-embarking. Those who actually went to the beach would often concur with many guidebooks that the physical experience was an anti-climax since the white turns out to be sharp gravel and the milky plumes show the dangerous rip currents that often stop bathing. In high summer the switchback road down to the beach is still often jammed full of parked cars. It seems an almost text book example of an image being promoted and then consumed by tourists.

And yet being a successful tourist is not so simple as this reading of the signs might imply. As people sat on the viewing platform, they could indeed marvel at the view – and who would not? They would also comment on being there to get 'the view' that they knew they had to have, with a degree of self-awareness that this was 'the picture' they were meant to take. As the guide on one party we travelled with put it, we did not need to worry because the bus would stop in exactly the best location to let you get 'that picture' – but please try not to get run over. An injunction to get the picture, with which they and others were largely happy to comply – save some anxieties felt by those without an authoritative guide that this was indeed the best point from which to do gain 'the picture'. Those that stopped at the viewing point would then look around and move off, possibly pausing only to walk round the spur to take an equally stunning view north towards Assos.

To situate this viewpoint we might look at the range of visualizations of the island. We can also turn to the humble resort map in the brochures which is helpfully outlined with parasols for major beaches and is then divided into lively and quiet resorts, which follow a roughly North–South divide. As the *Daily Telegraph* put it 'Sand is common, but pebbles keep things posh... [in the North there are] women in bikinis (no thongs here), sarongs and sandals' (*Daily Telegraph*, 18 August 2001). Myrtos is firmly in the Northern side of the divide. And of course this chimes with how Greece plays on the myth of the untouched Edenic beach (Lencek and Bosker 1998) in its publicity. Empty, and populated only by the occasional couple, the Greek beach of the posters and brochures offers a chance to 'live the myth' of romantic solitude – to adapt the GNTO 2005 campaign phrase.

By contrast the 'long' and 'wide' beaches (Makri Gyalos) of Lassi found their soft sand densely filled. Here the beach-based tourists congregated with beach bars and the like. Such resorts have been characterized as part of the European 'space that has become the most effective substitute for the time of the breaking-up party, that countryside festival that industrialization eliminated from the calendar of Europeans' where British tourists 'give free rein to their true desires, which are none other than making their holidays into a string of Friday nights in their pubs and Saturday mornings in the sunny Mediterranean [...] without there being the slightest hint of the pretension of getting to know other countries and other cultures' (MVRDV 2000: 107, 117). In this context Galani-Moutafi (2000: 210) notes somewhat acidly of tourist trends to Greece, that current tourists seek less cultural outlets and more stimulation of the senses – reflected in a tourism marketing imagery mix of sandy beaches, retsina, ouzo, bouzouki and syrtaki dance capped by the Parthenon. He suggests this is a postmodern tourism which reaches out to affect the audience through immediacy not aesthetic categories.

In the discourse of the brochure 'romantic' here stands in opposition to 'lively', with both being readable through a class sensibility. Thus Mytros, even in its high-season crowds, was set apart by the image – from so far back figures on the beach blurred – and by the need for cars to visit it and the shingle framing the beach. Here class seems writ not only in the aesthetics of looking at the beach but the corporeal ways of being upon the beach, through the proximal sensate world of being on the beach, inhabiting the beach as much as the distal world of visual imagery (Obrador Pons 2007). While it is usual to depict the classes as separated by the more bodily engagement of popular practice this is to misread the simply different haptic practices. It is not here a question of seeking the unspoilt as a marker of class distinction and taste, though that is reflected in the commentary – that seeks to position some as 'turistas vulgaris' (Löfgren 1999: 264) who travel in 'herds', 'stampede' onto beaches, 'flock' to see places, and 'swarm' around 'honey-pots' – nor is this simply a matter then of quests for the next untouched beach, despite the isolation promised in the imagery but often not delivered on the island (Crang 2004). In line with the book and the movie it is a more middle-brow aesthetic and affordance. More mixed and more fragmented. Thus there are more or less knowing performances of the obvious in terms of pictures taken. There is togetherness on beaches and then there are moments of quiet.

Losing Corelli and Evading the Cinematic Gaze

So how does one visit Corelli's Island through this visual and haptic economy? At one level one cannot avoid arriving there – tour reps have to watch the movie as part of their induction – and following the framing in the guides and brochures. And yet what does arriving there mean? One can find the island, one would think, on the bus tour helpfully labelled 'Corelli's tours'. And yet this is the same as most others labelled 'island tours'. What is perhaps remarkable is the absence of Corelli's island in two registers – both directly through the movie and secondly the island out of time the movie depicts. To begin with thinking about the former there is a deliberate relative absence of Corelli from the explicit tourist marketing *on* the island. Thus as one British resident writing a review for *The Times* (7 August 2007) commented in general: 'The film was good for Cephalonia. Although the critics panned it, it was a showcase for the island's scenery and boosted tourist numbers'. And in looking at recent marketing and development she concludes:

> the island that is the star of those enticing Greek travel posters… The response to the film of *Captain Corelli's Mandolin*, which was made here was a good yardstick of the Cephalonian mindset. On some islands the film would have taken over, but there is only one small restaurant near Sami called Captain Corelli's, and that's where the actors and film crew used to eat.

Indeed to be pedantic there are two such cafés though owned by the same person. Indeed his renaming them was the cause of debate. At first it was an example picked upon in many travel articles to suggest a Corelli-mania was sweeping the island in 2001–2. For the locals this notion of themed sites represented something of a trap, and one with which they wrestled then and since. Many worried that branding in this way was counter-productive, with a villa holidays specialist being quick to comment of her friend's Captain Corelli's café that she had warned him 'My clients would frankly avoid it' as kitsch and being 'too obvious'. Indeed, tourists too often singled out the Café for opprobrium with comments such as 'Captain Corelli's café – yeah too obviously a tourist trap'. And this resistance is shared by de Bernières himself

> A good friend of mine … who runs a cafe in Fiskardo, likes to tell me that I have ruined his island. He is only half serious, I hope, but it is a thing that worries me none the less. I was very displeased to see that a bar in Aghia Efimia has abandoned its perfectly good Greek name, and renamed itself 'Captain Corelli's', and I dread the idea that sooner or later there might be Captain Corelli Tours or Pelagia Apartments. I would hate it if Cephallonia were to become as bad as Corfu in places, with rashes of vile discotheques, and bad tavernas full of drunken Brits on two-week, swinish binges (de Bernières 2001: 15).

This resistance to marketing was embodied in the way the movie's local legacy was handled through the mayor of the small town of Sami. While many marvelled at the way Old Kefalonia was recreated, he had insisted that all physical traces of the movie should be disassembled and removed – to stop the area being damaged by the production. On-site commemoration consisted of to our certain knowledge 3 display boards, each some 2 metres by 1.5 metres comprising numerous 15 inch by 25 inch or so stills from the making of the movie and usefully emblazoned in Greek and English with 'Municipality of Sami'. That these boards were cryptic could be judged from the often baffled stance of tourists ambling by, who stopped peered, peered again then ambled away little the wiser. Otherwise the other main relics are indeed representational – with a local photographer in Sami selling stills of the making of the movie, which are syndicated throughout shops in the island, and posters or stills used by businesses, usually to celebrate their (purported) linkage to the making of the moving. Thus one hotel enterprise has signed pictures of Penelope Cruz and director John Madden, to celebrate how it was a base for the movie crew – though not it has to be said Cruz or Madden.

To find the relics of the movie producing Corelli's Island, was, then not a straightforward task. For the dedicated pilgrim one can find guides and books and specific features on the DVD to enable you to start looking. But without these sources there is nothing much there to guide you, and even with them there is little to confirm that the location is indeed correct. Thus the sequence of Pelagia dancing, which established her charm as well as staging Greek folk life as happy and full of spontaneous dance-led festivals, was shot using a deserted village threshing floor that it is possible to find – with a car, an hour or two of time and some determination. The main locations where the Greek village was built, or the Italian camp, or the capital city have no distinguishing material marks save the signboards. So while there are restaurants or shops that use the movie in billboards, most typically of Cage and Cruz in an embrace against the setting of the fishing jetty in a tranquil bay (an image from the cover of the edition of the book generally on sale in the island) there is little beyond that invocation or a romantic place. It mobilized, as perhaps nothing else on the island, the idea of the Romantic and secluded beach.

If we look a bit harder at that example of the fisherman's cove then we can start to see the process of erasure and active resistance to the branding of Corelli. The cove was on no movie maps. And even when equipped with detailed road maps and a fluent Greek speaker, we ended up in the wrong location – to be regaled by the local goatherd who lived there that they should have used his beach. The fisherman's cove setting is also the only place with a physical relic of the movie – the fishing jetty. That this survived was a testament to local government boundaries, it being in a different municipality than other sites. Having got lost we found the road sign to the beach on a back road through a small village. Indeed, we made out that it was the sign, despite the local graffiti that had obliterated most of the content and replaced it with directions to the new Athens airport. The locals did not wish, as the owners of the beach put it, 'dirty tourists' to mess up their local beach. Written through this, then, was a sense of local non-commercial use, but also in terms of location the tourists in question were not coach parties of mass tourists. They were tourists in cars, who might go camping, and were thus predominantly Italians over on the ferries. Meanwhile the municipality had not only kept the jetty, but even maintained it – and as a tourist attraction since it was 'fake' in that being a film set you could not actually moor boats on it. On this beach a handful of children (mostly from Italian families who had come by car ferry) played, and the occasional boat hired by other tourists was moored. Here, then, the politics of differential possession and mobility changed once more, between the mobile Northern European, the mobile Italian and what *The Times* (without irony) called 'Britannicus hedonisticus' (2 September 2000) that dominated resort beaches in the south and west in the imaginaries of many locals and tourists.

The second erasure of Corelli's island is the erasure of what it might signify about an authentic and 'old world' Greek society. If we might see that one thing beyond simply natural scenery that the movie portrays, then the world it evokes – a world out of time – is markedly absent. Studies on Symi (Damer 2004) and Serifos (Terkenli 1999) point to the role of the vernacular architecture in signifying an everyday Greekness to life. In the case of Serifos, then the urban fabric serves to encode a Greek way of life even as the traditional social activities associated with the built forms have been drained from the built up areas. On Symi the celebration of an atypical romantic architecture is used to mark the island off the coast of Turkey as quintessentially Greek – signifying its belonging in the Hellenic ethnos. But the tragedy of the earthquake in 1953, recorded vividly in the novel though not the film of Corelli, largely obliterated the distinctive Kefalonian architecture. Only two buildings were left standing in the capital, and now tours are run to the Northern village of Fiskardo that 'miraculously' survived. For the film, one drawback on the island was the absence of actual scenery to use – it all had to be built. One advantage is the large number of deserted villages and locations in the mountains, as after the earthquake villages were rebuilt on coastal sites. If tourists are drawn to the built forms of elegant Venetian-influenced townscapes they will be disappointed by their absence.

For the islanders themselves there is not quite the sense of continuous history, nor the competition for power through social memory in preserving built forms (Herzfeld 1991) or the selective preservation of elite dwellings (Damer 2004). There is a traumatic break in the fabric of memory. Indeed Dr. Ionides' focus in his *Brief History of Cephallonia* on the gales and tempests of events presages the traumatic times of occupation, civil war and natural disaster in the novel. The novel itself stirred unsettled and angry arguments on the island over the occupation and more so the civil war. De Bernières' depiction of singing and laughing Italians, versus the German iron military is not regarded as generally representative of the harsh times of the war. In terms of formal monuments to the period, the Italian occupiers, the Aqui division who were massacred, are commemorated by a small monument set outside the capital. The politics of memory are complicated and convoluted. In a study of Kalymnos, Sutton argued that such memories, traditions and histories played different roles in fashioning and forging notions of local identity (Sutton 1998). Our aim here is not an ethnography of local memory but rather a sense that the history here is both present yet physically absent. The film glossed over and the novel stirred troubled stories, which do not have a generally accepted, monovocal account. The period though speaks to an island before the fall – before the earthquake – so that for locals it is something of a land that time has forgotten. The scenography has become something of a hyperreal history where postcards in shops titled 'Old Argostoli, Kefalonia' actually show the movie set simulation set in the town of Sami.

The land of Greek folk custom and vernacular architecture though is scarcely evoked in contemporary tourist development. Many tourist-related business are wary of speaking of the earthquake – worrying seismic instability is not a great selling point for the island – and as noted the physical fabric is largely destroyed or ruinous, and often located away from the new villages that have become tourist centres. So there is a sense of lost world haunting the island but not one that speaks to many of the visitors. In the tourist village of Skala, perhaps the largest resort on the southern coast, there is a map showing the old village in a guide. It is a photocopied amateur booklet, buried amid the glossy materials on display in kiosks, drawn from memory and hardly a major device in shaping touristic practices.

So for all the clamour and excitement the island of Corelli is hardly there. People may come and seek it, and if they come they can hardly miss it, yet also it is curiously hard to actually find. For locals there is clearly a pride in the movie, both for its depiction of the island but rather than its account of their history the making of the movie is celebrated more as an event in itself in their history. Thus locals retain memorabilia and memories, and the pictures and posters seem as much a celebration of their part in the movie and the movie was of the island. It does perhaps enhance their already existing determination to preserve the natural scenery and avoid the large and visually intrusive commercial development found on some other islands. For locals it thus represents both a threat of commercialization as an opportunity for marketing the island for romantic tranquil tourism. It thus becomes a stake in the exchange between sand and shingle, mass and romantic tourism for the island. In this though the book perhaps mediates a sense that the result is not about elite versus mass but variants around the middle brow.

Conclusion

Captain Corelli and Kefalonia seem at first glance to have all the makings of the tightest of fits in movie-related tourism. But in the end Corelli's Island is a phantasm for tourists, locals and academics. For locals the phantasms are multiform, with the spectre of mass tourism often mentioned, while the turbulent past and the loss of the old villages remains for many. Using Derrida's spectral analysis, we could claim that Kefalonia lives through its own ruins where the present landscape is produced out of a process of construction through destruction, through multiple deaths (Wigley 1993: 43; Derrida 1994). There are the ruins from the earthquake haunting the present landscape with architectural traces of a past forever lost and then, there are the ruins of the filmic space representing a reinvented past that returns to the present landscape to haunt it with memories *in absentia*. 'Derrida's "spectral analysis", after all, involves the return of a ghost' (Crang and Travlou 2001: 174).

For visitors, the movie clearly plays off a number of tropes of the romantic island in general, rural folk and Greeks in particular. It fosters and creates a sense of an idyllic love story set out of time. This scripting of the island plays up a set of affordances that speak to the possibilities for a romantic tourist destination. Indeed wedding tourism has become a niche but growing market for the island. Is this directly attributable to the movie? It is hard to find direct links between visitor practice and the movie or book. It is quite hard to avoid indirect links, however. What we might find then is rather that the book and film are part of a currency where locals and others imagine a tourism appeal that engages not in the carnivalesque hedonism of many resorts but in what Jansson (2002) calls the imaginative hedonism that consists of sublated satisfactions in the symbolic realm, here in terms of romantic scenery. Corelli opens the possibility for people to feel and connect with a romantic sense of a different world, and a sympathetic sense of scenery. It creates an imagined *topos* but one read through the contemporary interests of tourism as much as the historic lens of the movie. People do not seek out the movie, but what the movie shows the island offers to tourists. In this sense the film makes Kefalonia stand out from among other Greek islands for offering those possibilities. The film does this rather than the book, since it offers the visualization of the scenery that potential tourists could not accomplish on their own. For academics, hunting for the example of tight linkage of movie, spectators and visitors like pilgrims then it is also phantasm. The influence is widespread yet modest. Far from being at the extreme end of the typologies of fans travelling to sites, it is more diffuse yet undeniably present in how many relate to the island. The island becomes a space of constructed visibility, one where the tourist visuality of beach and sand, mass tourism and romantic solitude overlaps only partly with a cinematic vision of myth and lost times. Despite the overcoding of book and film on the island, the local response has held a gap between the cinematic and touristic appropriation of the island.

References

Acland, C. 1998. Imax technology and the tourist gaze. *Cultural Studies*, 12(3), 429–445.

Busby, G. and J. Klug 2002. Movie-induced tourism: the challenge of measurement and other issues. *Journal of Vacation Marketing*, 7(4), 316–332.

Connell, J. 2005. Toddlers, tourism and Tobermory: destination marketing issues and television-induced tourism. *Tourism Management*, 26, 763–76.

Crang, M. 1997. Picturing practices: research through the tourist gaze. *Progress in Human Geography*, 21(3): 359–74.

Crang, M. 2003. Placing Jane Austen, displacing England: touring between book, history and nation, in *Jane Austen and Co. Remaking the Past in Contemporary Culture*, edited by S. Pucci and J. Thompson. New York: SUNY Press, 111–132.

Crang, M. 2004. Cultural geographies of tourism, in *A Companion to Tourism*, edited by A. Lew, A. Williams and C.M. Hall. Oxford: Blackwells, 74–84.

Damer, S. 2004. Signifying Symi: setting and performance on a Greek island. *Ethnography*, 5(2), 203–228.

Dann, G. 1996. The people of tourist brochures, in *The Tourist Image: Myths and Myth Making in Modern Tourism*, edited by T. Selwyn. Chichester: Wiley, 61–82.

de Bernières, L. 2001. Introduction, in *Captain Corelli's Mandolin: The Illustrated Film Companion*, edited by S. Clark. London: Headline Books, 9–15.

Derrida, J. 1994. *Spectres of Marx: The State of Debt, the Work of Mourning and the New International*. London: Routledge.

Galani-Moutafi, V. 2000. The self and the other: traveller, ethnographer, tourist. *Annals of Tourism Research*, 27(1), 203–24.

Herzfeld, M. 1991. *A Place in History : Social and Monumental Time in a Cretan Town*. Princeton, NJ: Princeton University Press.

Hudson, S. and Ritchie, J. 2006a. Film tourism and destination marketing: the case of Captain Corelli's Mandolin. *Journal of Vacation Marketing*, 12(3), 256–68.

Hudson, S. and Ritchie, J. 2006b. Promoting destinations via film tourism: an empirical identification of supporting marketing initiatives. *Journal of Travel Research*, 44, 387–396.

Jansson, A. 2002. Spatial phantasmagoria: the mediatization of tourism experience. *European Journal of Communication*, 17(4), 429–43.

Kim, H. and Richardson, S.L. 2003. Motion picture impacts on destination images. *Annals of Tourism Research*, 30(1), 216–37.

Lencek, L. and Bosker, G. 1998. *The Beach: The History of Paradise on Earth*. London: Secker & Warburg.

Löfgren, O. 1999. *On Holiday: A History of Vacationing*. Berkeley, CA: University of California Press.

Macionis, N. 2004. *Understanding the Film-Induced Tourist*. International Tourism and Media Conference Proceedings. Melbourne, Monash University.

MVRDV 2000. *Costa Iberica*. Barcelona: Actar.

Norris Nicholson, H. 2006. Through the Balkan states: home movies as travel texts and tourism histories in the Mediterranean, c.1923–39. *Tourist Studies*, 6(1), 13–36.

Obrador Pons, P. 2007. A haptic geography of the beach: naked bodies, vision and touch. *Social & Cultural Geography*, 8(1), 123–141.

Papadimitrou, L. 2000. Travelling on screen: tourism and the Greek film musical. *Journal of Modern Greek Studies*, 18, 95–104.

Riley, R., Baker D. and Van Doren, C. 1998. Movie-induced tourism. *Annals of Tourism Research*, 25(4), 919–35.

Sheppard, R. 2002. Savagery, salvage, slaves and salvation: the historico-theological debate of Captain Corelli's Mandolin. *Journal of European Studies*, 32(1), 51–61.

Sutton, D. 1998. *Memories Cast in Stone: The Relevance of the Past in Everyday Life*. Oxford: Berg.

Terkenli, T. S. 1999. Livadi, Serifos, place identity, in *Tourism and the Environment: Regional, Economic, Cultural, and Policy Issues*, edited by H. Briassoulis and J. van der Straaten. Dordrecht: Springer, 179–203.

Tooke, N. and Baker, M. 1996. Seeing is believing: the effect of film on visitor numbers to screened locations. *Tourism Management*, 17(2), 87–94.

Tzanelli, R. 2003a. 'Casting' the Neohellenic 'Other': Tourism, the Culture Industry and Contemporary Orientalism in 'Captain Corelli's Mandolin' (2001). *Journal of Consumer Culture*, 3(2), 217–44.

Tzanelli, R. 2003b. 'Disciplining' the Neohellenic character: records of Anglo–Greek encounters and the development of ethnological historical discourse. *History of the Human Sciences*, 16(3), 21–50.

Urry, J. 1990. *The Tourist Gaze: Leisure and Travel in Contemporary Societies*. London: Sage.

Wigley, M. 1993. *The Architecture of Deconstruction: Derrida's Haunt*. Cambridge, MA: MIT Press.

Chapter 5

The Mediterranean Pool: Cultivating Hospitality in the Coastal Hotel

Pau Obrador Pons

Introduction

As the main provider of hospitality to mass tourism, the coastal hotel occupies a prominent position – both physically and symbolically – on the coastlines of the Mediteraranean. The coastal hotel is a typical space of our times, which as with the case of the airport (Gottdiener 2001) and the shopping mall (Goss 1993), responds to the abstract principles of movement and commodification. It is a smooth and fluid site, notorious for its homogeneity and lack of character, a fleeting environment where people come and go and which does not encompass the stability and endurance traditionally associated with place and community. Although in recent years new models have emerged, most coastal hotels are still built according to an standardized formula: three or four stars, relatively close to the beach, a big pool at the centre, exuberant vegetation and an architectural style that was modernist in the 1960s and now is increasingly more pastiche. It may be a highly commodified space, but it is one in which millions of tourists have fun, relax and socialize every summer. Coastal hotels, and in particular their pools, are theatres of sociality and distraction that sit right in the middle of the Mediterranean tourist experience.

The significance of the coastal hotel as a tourist experience has been consistently overlooked. Despite being one of the prime tourist spaces in the Mediterranean there are very few examples of research that examine what is actually going on by the hotel pool, the work of Andrews (2000, 2006) being a rare exception. When the coastal hotel features in tourism literature, it does so mostly in the context of relations of production (Buhalis 2000) as an example of neofordism (Ioannides and Debbage 1998) or as a context for exploring the significance of gender in tourist employment (Sinclair 1997). Tourism literature is consistent with the social sciences' view of the hotel lobby, which emphasizes the inhabitability of contemporary spatialities. While bringing to light the economic principles underpinning mass tourism, tourism literature effectively 'follows the logic of the corpse' (Thrift 2004: 83), draining spaces of hospitality of meaning, emotion and social relations. Similarly most hospitality literature focuses exclusively on the economic order of things, undervaluing the everyday relationships and practices through which the coastal hotel is consumed. In the work of Buhalis, Sinclair and

Ioannides and Debbage, little attention is paid to the actual experience of staying
in a hotel, the mundane forms of hospitality that underpin the economies of mass
tourism. In the coastal hotel we are confronted with the inability of the social
sciences to make sense of contemporary spaces, ephemeral experiences and the
banal.

This chapter contrasts general statements about the inhabitability of
contemporary spatialities with an ethnographic account of day to day uses of
the coastal hotel. Re-examining theoretical statements in the light of ethnographic
material in this way allows us to break with the impasse of dominant perspectives,
thus engaging more productively with the actual tourist experience of the
Mediterranean. Taking my research, conducted on site at two hotel pools, as my case
study, the chapter draws attention to the pressures and pleasures of sociality which
inhabit the coastal hotel. The focus of this chapter is on the fleeting and mundane
moments of hospitality, the network of performances of hosting and guesting that
makes the coastal hotel livable. I am interested in how hospitality and sociality
are organized and made meaningful by the pool, dwelling in particular in the role
of commercial hospitality in promoting and producing convivial economies and
a sense of being together. This chapter therefore reflects on the nature of social
relations between people in the highly commodified and fleeting environments
of mass tourism where the categories of host and guest are uncertain. Inspired
by recent discussions on commercial spaces of food and drink (Bell 2007a,
2007b, Latham 2003, Mozl and Gibson 2007), that draw on Derridian notions
of hospitality (Derrida 2000, Deutscher 2005), the chapter moves away from the
social sciences' emphasis on the inhabitability of contemporary spatialities to
consider the coastal hotel in terms of hospitality. Ideas of hospitality offer the
possibility of addressing the sociality of the coastal hotel without reverting to
notions of organic community.

This chapter is organized in three sections. The first section examines
dominant perspectives on contemporary spatialities, which tend to emphasize the
inhabitability of the generic landscapes of late capitalism, and their suitability
for the study of mass tourism. The second section develops an alternative
understanding based on Derridian notions of hospitality and Maffesoli's analysis
of contemporary cultural life, opening up mass tourism to new forms of analysis
which are less hostile to this tourist phenomenon (1996). The chapter finishes with
an ethnographic analysis of practices of hospitality and inhabitability in the hotel
pool, the main communal area in the coastal hotel. This ethnographic approach
allows me to contrast the important day to day experiences of being in the coastal
hotel with the discourse that would see these places as lifeless and inauthentic.
This chapter is based on my doctoral research that I carried out in two large hotels
in the island of Menorca (Spain). Both hotels are paradigmatic of 'sun, sea and
sand' tourism, which in Menorca is predominantly family orientated. The research
combined participant observation and serial interviews with groups of British
tourists. I spent a total of three weeks in each hotel, interviewing 70 people in total.
In order to produce a full account of their holidays, each group was interviewed

once every three or four days for a period of between 20 minutes and one hour. The respondents were also asked to fill in time diaries. During my time in the hotel I also participated in a number of entertainment activities and carried out systematic description of the communal areas.

The Inhabitability of Contemporary Spaces

The coastal hotel is paradigmatic of the landscape of late capitalism, a smooth, nomadic, frictionless site that responds to the principles of movement and commodification. The proliferation of generic landscapes has been a major concern of contemporary cultural theory. The question occupies a prominent position in the work of the twentieth century critics of space and everyday life including Benjamin, Simmel, Kracauer, Foucault, de Certeau, Augé and Debord. However, the dominant view has been primarily negative. Most critics are deeply suspicious of the proliferation of highly commodified and fluid spaces such as the coastal hotel. Contemporary spatialities have been conceived mainly in terms of (in)habitability and failure, re-enacting a nostalgia for organic communities. As Buchanan and Lambert explain 'at stake is the practical problem of what it takes to make space habitable, to make places from sites where the active place making infrastructure (tradition, memory, habit and so forth) had been either destroyed or displace' (2005: 2). There is a general conviction that contemporary spatialities are at best highly ambivalent and very often essentially unlivable. This section considers to what extent the social sciences' emphasis on the inhabitability of contemporary spatialities provides a compelling basis for analysing the tourist realities of the coastal hotel.

There is a long tradition of thought that addresses the changing nature of contemporary spatialities. Simmel's conceptualization of the stranger (1971) together with Benjamin's work on arcades, are two of the earliest and most influential analyses. The figure of the stranger, that for Simmel embodies the contradictory experience of the modern everyday (Allen 2000), speaks of the difficulties of dwelling in the contemporary fluid spaces of the metropolis. His take on contemporary spatialities is marked by a fundamental ambivalence. While seduced by the freedom and excitement of the modern city, the metropolis is identified as the breeding ground of all sorts of diseases attributed to spatial conditions. For Simmel, the disease is the blasé, a modern attitude of indifference and insubstantiality. Modern life, with its sharp intensification of nervous stimulations, results in a feeling of loneliness and indifference that impregnates human existence (Simmel 1997). While Simmel addresses the modern condition of estrangement as well as the changing relations of proximity and distance in the metropoli, Buchanan and Lambert point out he is not yet 'prepared to consider the possibility of spaces that are constitutively uninhabitable' (2005: 3); in Simmel estrangement goes hand in hand with freedom. There is no such hesitation in the work of Foucault, de Certeau, Debord and in particular Augé, where the generic landscapes of late capitalism are dismissed as unhomely and empty (Buchanan and Lambert 2005: 3). Augé (1995) speaks of a contemporary super modernity multiplying uninhabitable spatialities like airports, highways and petrol stations (Augé 2000), a permanent elsewhere 'in which people are always and never at home' (1995: 109). His work relies on a distinction between *place*, which can be defined as 'relational, historical and concerned with identity' (1995: 77) and *non-place*, which fails to confer the affect of place, creating instead solitary individualities. Non-places, which for Augé would include the coastal hotel, are smooth, streamlined spaces surrounded by the fleeting, the temporary and the ephemeral in which abstract and contractual relations prevail over organic solidarities.

Writing in the 1930s, Sigfried Kracauer translates this suspicion of contemporary spatialities to the hotel lobby, while maintaining a sense of ambiguity. For Kracauer the hotel lobby is a spatial metaphor of the emptiness and homelessness that characterize modernity (Katz 1999, Tallack 2002, Mülder-Bach 1997, Reeh 1998). The hotel is identified as the inverted image of the house of god, as 'the space that does not refer beyond itself' (1995: 176–7). It is not a site of meaning, transcendence or organic community but of superficiality, void and fragmentation. 'Whereas in the house of God a creature emerges which sees itself as a supporter of the community, in the hotel lobby what emerges is the inessential foundation at the basis of rational socialization. It approaches the nothing and takes shape by analogy with the abstract and formal *universal concepts* through which thinking (…) believes can grasp the world' (1995: 179). In the hotel there is no meaning or purpose in being together. Hotels are sites of pure coincidences where people and events have no other relation to each other than the fact that they happen to be in the same place at the same time. Like the non-place, the hotel is a spatial desert, trackless and depthless, the quintessential nomadic space, smooth and open to traffic. Its depthlessness and abstract character makes the hotel lobby 'a herbarium of pure externality', a site of pure distraction and mere play 'an aimless lounging, to which no call is addressed, leads to the mere play that elevates the unserious everyday to the level of the serious' (Kracauer 1995: 179).

The tourist realities of the coastal hotel sit uneasily with this tradition of thought. The popularity of the coastal hotel poses an intellectual challenge to the social sciences' emphasis on alienation and commodification, in particular to the idea that commodified spaces inevitably lead to commodified experiences. Dominant perspectives tend to ignore the specificity, multiplicity and locatedness of all human engagement with the world. Arguments for the impossibility of dwelling in contemporary spaces depend on an undervalorization of the everyday social relations and practices these spaces afford. The work of Kracauer (1995) and Augé (1995) show little interest in how individuals and groups go about building a world in highly commodified spaces, insisting instead on an abstract 'ratio' of capitalism shadowing the creativity and spontaneity of the subject. 'Individuals are nothing more than members of an audience' – Hetherington explains – 'they exist outside of their own bodies. They cannot encounter the world other than as isolated and static members of an audience' (1998: 61). In emphasizing the inhabitability of contemporary spatialities, dominant perspectives reproduce the values of high art, in particular an anxiety over massification. A fear of cultural alienation and massification are persistent themes in the social sciences and illustrate a celebration of bourgeois aesthetic principles. For most critics, 'the heroic bourgeois *Persönlichkeit* self-reflexive and motivated by commitment to the rigors of a self-disciplining vocation, is taken as the ideal personality' (Hetherington 1998: 46). Cultural forms such as the coastal hotel that depart from this individualistic self-reflexive ideal are seen as a sign of moral decline.

To make sense of the coastal hotel it is necessary to engage more positively with the particular kinds of spaces and social practices associated with mass tourism. We need to move beyond notions of commodification and alienation to examine how tourists actually inhabit and make space in the costal hotel. Rather than adopting Augé's point of view, this chapter takes as its starting point the conclusions of Crang (2002) on how to think of interstitial places and ephemeral moments. Crang concludes that 'they are not just liminal or threshold spaces, though there is that element. Nor are they simply places of homogenized, commodified experience; nor just the rationality of scheduling and "flow management", through they rely upon both. The few examples above suggest they are also places of fantasy and desire, places of inclusion and exclusion, and social milieux for different groups of people'. A few lines later he reiterates 'They are not places where people are at home, though they are familiar. They are not just the absence of "organic community" but offer different pleasures and pressures of sociality when the situation is ephemeral from the start' (Crang 2002: 573). The coastal hotel is not simply a place of liminaliy, nor of commodifiation or the disintegration of organic communities. The hotel, like the swimming pool (Molotch 2000) or the bus (Hutchinson 2000) is also a site of relevant social and cultural practices that speak to identity, fantasy, sociality and exclusion. It is these social and cultural practices, 'these pleasures and pressures of sociality', which form the basis of this chapter. Before I examine them in more detail however I would like to sketch out a different perspective of the coastal hotel, drawing on recent discussions of spaces of hospitality (Bell 2007a, 2007b, Molz and Gibson 2007) and postmodern socialities (Maffesoli 1996).

From Inhabitability to Hospitality

In this section I consider the coastal hotel in terms of hospitality rather than inhabitability in an attempt to open up mass tourism to new forms of analysis which are less hostile to it as a tourist phenomenon. The main appeal of the idea of hospitality is the possibility of thinking the sociality of the coastal hotel without recurring to notions of organic community, alienation or commodification – as Bell explains, 'it offers ways of being-with-others which are inaccessible through community' (2007a: 10). Philosophies of hospitality have found their way into the social sciences through the work of Jacques Derrida (2000) and its application by a number of social scientists (Deutscher 2005, Molz and Gibson 2007). Derrida's work on this issue has been highly influential, 'reinvigorating theoretical and philosophical debates about ways of relating, about host and guest' (Bell 2007a: 9). The work of Derrida on hospitality has to be understood in the context of his discomfort with ideas of community and cultural identity; in particular with the politics of self-other relations that opposes the identity and belongingness of the host against the movement, shifting and instability of the guest. 'For Derrida' – Caputo explains – 'one must watch out for the ways tradition and community become excuses for conservatism, for the exclusion of the incoming of the other, and hence, constitute as much threat as promise, as much a trap as a tap' (1997: 109). Rather than emphasizing closeness and similarity, the Derridian approach to hospitality opens up ways of thinking about difference, inviting and welcoming the other. 'Derrida reacts not by denouncing the very idea of cultural identity, but by deconstructing it, which does not mean laying it to the ground or leaving it in shambles but opening it to difference' (Caputo 1997: 114). In examining the nuanced fluidity of the categories of host and guest in mass tourism this chapter embraces Derrida's critique of the notion of hospitality as well as his demand to open it up.

Notions of hospitality allow us to engage more positively with the pleasures and pressures of sociality in the coastal hotel, unblocking the impasse of dominant perspectives generally oblivious to the life that the generic landscapes of late capitalism afford. Rather than emphasizing the disappearance of community, Derridian ideas provide an opportunity to address the significance of mundane forms of hospitality and an ethics of conviviality in the composition of contemporary spatialities. There is more to the coastal hotel than commodification and alienation. They are theatres of sociality animated by a series of hosting and guesting performances as well as a variety of social encounters and modalities. Its success depends upon the mobilization of certain ways of inhabiting and being together that transgress the stability associated with organic communities. The enactment of a welcoming feel is critical for the coastal hotel. Derridian notions of hospitality shift the question from the possibility of social life to its differential organization in a fleeting environment full of strangers that come and go. Recent discussions have explored the potential of bringing together philosophies of hospitality and hospitality studies. The edited collection edited by Molz and Gibson (2007) and Bell's progress report (2007a) are the latest and most explicit attempts. Focusing on spaces of commercial food and drink, Bell looks at how certain version of hospitality are woven into urban regeneration schemes. As well as responding to commercial imperatives, the new bars and restaurants emerging in the city play a vital role in the production of new urban identities and modes of living in the city. They are an important part of making 'the hospitable city' (2007a: 8). The link between the processes of urban regeneration, commercial hospitality and the mobilization of hospitality also forms the basis of Latham's work on a gentrifying neighbourhood in Auckland, New Zealand. For Latham 'what is happening there is about more than an aesthetic of consumption' (2003: 1706). The new cafés and bars that have emerged in Ponsby Road have acted as 'a key conduit for a new style of inhabiting the city' (2003: 1710). A new polymorphous public culture has been organized around spaces of consumption. 'Consumption has quite literally helped to build a new world' (2003: 1713). As with the case of the new spaces of food and drink, the coastal hotel cannot be reduced to narrow economic relations. A typical site of our time, it also speaks a language of culture, identity and sociality in novel ways.

The dominant forms of sociality in the coastal hotel consist of fleeting and mundane experiences mostly dedicated to the pursuit of fun. The kind of sociality coastal hotels afford are in most cases insubstantial and do not rely on a communal project or ideology but on warm companionship and the physicality of being together and having fun. The crucial point is the act of creation as such, the puissance of everyday life. Maffesoli's analysis of contemporary cultural life is conceptually useful for thinking these 'light touch forms of sociality' (Thrift 2005). It provides us with a description of the dominant forms of hospitality in contemporary spatialities. The main virtue of Maffesoli's approach is that he centres his analysis of contemporary cultural life in the expressive realm of feeling and emotion (Hetherington 1998: 53). Maffesoli speaks of the rise of neo-tribalism, a new form of sociality which is affective, elective and tribal in character. There is a new kind of 'holism taking shape before our eyes: throwing wide the doors of privacy, sentiment takes over, and in certain countries its presence is reinforced in the public sphere, thus producing a form of solidarity that can no longer be denied' (1996: 16). The emphasis of Maffesoli is on the flickering practices of hospitality in a mobile world. In contrast with the stability induced by classical tribalism, neo-tribalism is a kind of gathering that is fluid and ephemeral. What binds together these contemporary groupings is not a rational project or an ideology but simply an ambience, a lifestyle or an emotion. 'We are witnessing the tendency for a rationalized social to be replaced by an empathetic sociality, which is expressed by a succession of ambiances, feelings and emotions' (1996: 11). The solidarity, the holism, that binds people together in contemporary neo-tribes does not derive from the institutional world of politics but from the puissance of everyday life, the expressive life of the individual. What is important is the moment, the execution of being together and creation for its own sake. In contemporary groupings solidarity is expressed in the mundane everyday through quite trivial elements. As Maffesoli points out 'having a few drinks, chatting with friends; the anodyne conversations punctuating everyday life enable an exteriorization of the self and thus create the specific aura which binds us together within tribalism' (1996: 25). The fleeing environment of the coastal hotel responds well to Maffesoli's analysis of contemporary cultural life with its emphasis in conviviality and fun.

The Pool

Nowhere in the Mediterranean resort are the 'light touch forms of sociality' underpinning mass tourism so evident as they are in the hotel pool. As such, it is a paradigmatic example of the significance of mundane forms of hospitality and of an ethics of conviviality in mass tourism. The pool is the main communal area of the coastal hotel and an important part of the Mediterranean experience. A large proportion of tourists spend their holiday time, up to six hours a day, lazing by the pool, sunbathing, reading, chatting and swimming. Contrasting sharply with the romantic excitement and sublimity of the beach, the pool is a much more down to earth social space, generally associated with notions of immediacy, informality and simplicity. As in the case of the café (Laurier and Philo 2004) the way of inhabiting and being together by the pool is marked by a sense of the mundane and the banal. The hotel pool cannot be reduced to a matter of commodification and alienation. It is first and foremost a space of conviviality, hedonism and relaxation, the success of which depends upon the mobilization of ephemeral and playful ways of inhabiting and being together. The pool is the main stage for the enactment of the warmly and welcoming atmosphere that identifies Mediterranean mass tourism.

Drawing on Derridian notions of hospitality (Derrida 2000) and Maffesoli's affective socialities (1996), this section develops an ethnographic account of this mundane social life in two hotel pools on the island of Menorca. The focus is on how people create a living space and relate with others by the pool as well as on the social forms and structures of feeling that this environment contains. This sense of the mundane incorporates three different dimensions (Binnie et al. 2007). Firstly, the informal and easy feel of the pool. The hotel pool is fundamentally a space for doing nothing other than sitting on a chair or lying on a sun bed, reading a thriller, filling in a crossword or floating on a lilo. Mundanity here conveys a 'negative' experience of relaxation, liberation and retreat from the constraints of public life. There is no consistent utopia associated with the pool that can provide a glimpse of an alternative life. Secondly, the hotel pool emphasizes domesticity. As well as the fluidity and insubstantiality of neo-tribalism, in the hotel pool we find the enactment of workable utopias related with familial and national identities. The pool is a world of families, children and play, a site of enduring social forms. Thirdly, the excessive character of the mundane complicates the picture further. A retreat from public life, the pool also functions as an space for the cultivation of mundane skills and sensibilities. The pool offers the possibility to explore the limits of hedonism and learn the difficult art of splashing out, showing off and doing nothing. As well as a gap, the pool is a space and a practice of subjectification, a site of becoming, at the heart of which there are often dreams of luxury. These three aspects of the mundane are explored in this section through a description of various elements that make up the social fabric of the pool including conviviality, domesticity, hostility, boredom and the gaze.

Figure 5.1   Drinking game by the pool

Conviviality

The pool is a theatre of sociality, the success of which depends upon forging a friendly and warm atmosphere in a context where people do not necessarily know each other. Meeting other tourists, having a laugh with friends over a drink, and playing with your children in the water are some of the main pleasures that tourists find by the pool. A sense of conviviality is intrinsic to the design of the pool, a communal space that sits at the centre of the coastal hotel and brings people together for prolonged periods of time. The pool, with its distinctive easy-going feel, affords plenty of moments of hospitality and friendliness, while lying on sun beds, queuing at the bar or playing with the children. Being-by-the-pool is about enjoying the company of others in an informal manner, be they friends, family or strangers. 'Light touch forms of sociality' are prevalent by the pool. Rather than a place of lasting relations and profound conversations the pool is one of occasional gatherings, spontaneous chats and fleeting encounters. The social life of the pool is a good example of the kind of sociality that Maffesoli (1996) associates with neotribalism. It is an ephemeral and fluid kind of sociality that does not rely on an ideological project, but on 'warm companionship' as well as the physicality of being together. The prevalence of 'light touch forms of sociality' was evident in my ethnographic research on coastal hotels. Most of my respondents identified a friendly and easy atmosphere as one of the main attractions of the pool. Alison from Sussex, for example confirmed that 'we stay here because well they [the children] do like it and it's easier. But there is more chance to meet English people and to make friends with them'. The mobilization of hospitality is central to the workings of pool.

The convivial character of the pool is actively engineered through organized entertainment. Coastal hotels usually offer an activity programme that covers day an evening and is aimed at both at adults and children. The primary aim is to create a warm and friendly holiday environment. The entertainment programme is a conscious effort to mobilize 'light touch forms of sociality'. Take for example the drinking game pictured in Figure 5.1. At 1pm every day the entertainment team appear 'by surprise' in the pool announcing a drinking game. Wearing fancy dress, in this case skintight skeleton suits and green masks, they go around the pool area making lots of noise and encouraging people to participate. Once there are enough participants the game properly starts. It is normally a very simple game, in this

Figure 5.2   A family posing for a picture with the entertainment team on the last day of their holiday

case a dice version of pontoon, Using a dice, the participants have to get as close as possible to seven and a half without going too far. The holidaymaker who gets the closest wins a free cocktail that she/he has to drink all at once standing on a chair in front of the public. The event is short, funny and uncompromising; it is about laughing and making fun. Using circus techniques, the entertainment team is able to create a festive and friendly atmosphere. The key to its success is not the game itself but the fact that the activity leaders are acting like clowns, making jokes, being funny and spontaneous. What is important here is the ambience of fun as well as the physicality of being together. Drinking games are good examples of the affectual kind of socialities that Maffesoli (1996) identifies with neotribalism.

Domesticity

Coastal hotels and in particular their pools are 'theatres of sociality' that rely on the mobilization of mundane forms of hospitality as well as an ethics of conviviality. The most important of these forms is the family. The mundanity of the pool and other spaces of mass tourism convey domesticity as well as relaxation. With the significant exceptions of places like Magaluf and Faliraki, Mediterranean mass tourism is a world of families, children and play. Pleasure is not gained in sightseeing but in the intimacy of the family and the exposure of the body to the sun, sea and sand. By the pool 'light touch forms of sociality' coexists with more enduring social forms that combine an ideological and an affectual dimension. Alongside the fluidity and insubstantiality which is paradigmatic of neo-tribalism, we find the enactment of workable utopias related with the family and the everyday.

At least in Menorca, mass tourism is conceived to a large extent for the enjoyment of families with children. Nowhere is this more evident than at the hotel pool, which is heavily inscribed with meanings of domesticity. The pool is quintessentially a space for children that affords multiples possibilities for enjoyment, experimentation and friendship. Not surprisingly, many families keep their children happy by spending all day by the pool. This was the case for the Peterson family from Essex. Although their hotel was less than 100 metres from the beach, they did not visit it until the fifth day of their holiday. Instead, they spent most of their time, up to six hours a day, by the hotel pool. The attraction of the pool for the children was the main justification, 'kids love the pool, they've never had enough of it' – Sally explained. On holiday the happiness and welfare of the children is a priority, 'if the children are happy and have fun, we are happy' – Sally concluded. In contrast to adults, children do not spent their time lying on a sun bed reading, but in and by the pool playing, running, jumping, diving and snorkelling. For children, the pool is not a passive space but an environment in which to have fun and make friends, sometimes despite language barriers. It is also the space where parents are more likely to play with their children. Adults find in the pool a rare license to play and be a child again. The significant presence of children is critical in perpetuating the playful character of the pool, which binds people together.

The pool is also a site of everyday politics and identity formation, paraphrasing Löfgren (1999) 'a cultural laboratory'; it is a site 'in which people derive pleasure from performing and narrating alternative identities and ways of being together' (Haldrup and Larsen 2003: 24). The pool is a good example of the familial and mundane dimension of tourism that, according to Haldrup and Larsen, tourist studies have struggled to account for. A longing for the simplicity and tenderness of family life underpins coastal tourism, which is often conceived in terms of return. By the pool, families are on their way to childhood, to a fixed version of family life that no longer exists (Löfgren 1999: 149, Chambers, 2003: 101–7). 'Families'– Haldrup and Larsen explain – 'are in constant need of performing acts and narratives that provide sense, stability and love to their familial relations. The more family life becomes fluid and based on choices and emotions, the greater the task of tourist photography to produce accounts of timeless and fixed love' (2003: 26). Most of the comments that I collected evoke a return to a happy, stable and loving family. We can see this in Linda's remarks that 'it's just really nice, we are all sweet together here'; or when Claire congratulates her family for being all together on holiday and 'getting closer'; or when Tim cites the summer holidays as an opportunity to spend time with his children. Tourists not only present their families as tales of success and love, they also act as such, playing with children on the beach, avoiding frictions or sharing the same room in the hotel. These narratives of the family respond to dominant mythologies of the family as the natural basic institution, paraphrasing Hirsch, as 'a spiritual assembly based on moral values' whose bond 'stems from instinct and passion' (1999: xv). Embedded in the materialities of the pool we find unambiguous notions of what families are meant to do, think and convey to others.

The pool then is a site of politics about the family and the everyday, that is, about the *private*. However, the identities that are being formed and performed by the hotel pool are right in the middle of a battle about 'class taste' and 'moral correctness', that is, a battle about the *public*. Holidays are presented as a travel inwards, to the feminized private life of the family, as a search for authenticity amongst themselves, as a retreat from the hard realities of male dominated public live. However, there is no privacy for the family in their search for the ideal domestic life. Children and parents will share the same room and they will be relaxing by the pool under the close inspection of other families' gazes. While it responds to notions of retreat, taking the family to a coastal hotel in fact involves the exposure of the family to the anonymous but sanitized public.

The Gaze

One major attraction of the pool is the possibility of looking at other tourists and in particular their bodies. Although looking at others is in many contexts socially unacceptable especially when it is done in an obvious manner, the pool provides a license to observe others. Whereas on the street people tend to avoid each other's gaze, by the pool people feel more willing to look and be looked at. The physical organization of the pool area with all the sun beds facing the pool reinforces the idea that people are on show. The pool affords what Anderson calls 'practical and expressive folk ethnographies' (Bell 2007a: 16). Like the case of the café and the market, it is a space in which it is acceptable to watch and observe. I agree with Molotch when he contends, 'part of the intrigue of the pool is the chance to see other people's bodies', and concludes, 'they hold more than water and expose more than the skin' (2000: 189). The hotel pool would lose its charm if it did not afford the possibility of looking at the strange and literally foreign bodies of tourists from other countries.

Although deeply embedded in the culture of the pool, very few people in my research were actually ready to accept the pleasures of the gaze. Linda from Bristol was one of the few. She identified sitting by the pool, drinking and watching other people as one of the main pleasures of the holidays. She would 'Just sit, sit and drink and watch people. Just watch. I'm quite happy to stand here, there is so much going on!' She pictured the hotel as an exciting place full of strangers to look at and gossip about. 'I like watching people' – she recognizes – 'especially in this hotel. There are so many different nationalities. At breakfast time, for example, you can see from what they are eating what country they are from'. José Lucas, a holidaymaker in his 30s from Valladolid also explicitly recognized the attraction of the gaze. He was not ashamed to identify 'watching people', which he describes as 'the Spanish national sport', as one of the main pleasures of being on holiday 'I would sit now in the reception to see everybody passing through', he admitted. He misses a good marina like the one in Benidorm to stroll around in the evenings and watch other people passing through.

The relevance of the gaze together with the significance of gossiping complicates the social life of the pool. As well as the mobilization of mundane forms of hospitality, the pool is a place for the enactment of social divisions, national rivalries and cultural differences. The gaze forms the basis of these processes of differentiation. Observable variations in style, behaviour and manners often serve to reinforce the barriers that differentiate the community of hotel dwellers along national lines. In the next section I explore in more detail the significance of national clichés through the example of sun beds. In the coastal hotel, national clichés are critical in both dealing with the other and neutralizing the uncomfortable questioning that the coexistence with foreign people brings about. Different processes of reterritorialization continuously reconfigure the open and volatile character of the pool.

Hostility

To be a perfect host is one of the main aspirations of any hotel. According to Bell 'to be a perfect host is to offer hospitality unconditionally, unreservedly, unendingly' (2007a: 9). While an ideal of pure hospitality may be the driving principle of the hotel, Derrida reminds us that the enactment of hospitality is never unreservedly free. There are conditions through which hospitality is given and received. The principle of welcoming the unknown anonymous other – the stranger and the foreigner – is constantly corrupted with traces of hostility. This paradox is not exclusive to the hospitality industry. At the heart of Derrida's philosophy of hospitality there is a similar paradox between an ideal hospitality which is by definition unconditional and the enactment of hospitality which is always conditional (Deutscher 2005: 67). This conditionality of enacted hospitality is evident in the hotel's latent elements of hostility. The most obvious element is price. To be accepted as a guest in a hotel you need to pay an established rate. It is not a gift but a purchase. Moreover, in the coastal hotel there are plenty of rules of etiquette and reciprocity that determine who is welcome and on what terms. These include dress code, noise rules and the subtle monitoring of everyday behaviour through the entertainment program. The deployment of hostility towards deviant behaviour is necessary to make sure that the coastal hotel is habitable for the 'right' kind of people, normally white middle class families.

One of the most characteristic settings for hostility are disputes over sun beds. It is very common, in particular among British tourists, to complain about empty sun beds which are permanently reserved. The sun bed story is very much a territorial conflict between British and German tourists for the physical and symbolic possession of the central space of the hotel. The basic idea is that the Germans put their towels on the sun beds very early in the morning although they are not used until later in the day. Take for example the comments made by Craig, from London. 'You say, I'm going to get a sun bed now, but they [the Germans] have arrived at five o'clock in the morning and put the towels on the tops'. The situation was not exclusive to this particular hotel. 'Everywhere you go on holiday' – he remarked – 'you can come at eight o'clock and all the sun beds are taken (…) but people don't like to lie on the sun beds until quarter to twelve'. Craig didn't blame only the Germans, he also reproached the hotel management for being inefficient and inattentive. 'What I've found odd' – he concluded – 'is that they create a remarkable bureaucracy for food and drink with tickets and so on but they haven't got anything for the sun beds. Nothing happens at all with the sun beds'. So this is not simply a conflict about the use of scarce resources; it is also a backdrop for the reenactment of nationalist clichés and prejudices about Germans and Spanish. Craig wryly uses the story of sun beds to illustrate how possessive the Germans are, how inefficient the Spanish are and how unfair the world is for the English. In the coastal hotel, the mobilization of mundane forms of hospitality coexists with processes of reterritorialization that reenact national rivalries, social divisions and cultural differences.

Boredom

An ideal of relaxation is at the heart of the informal and mundane atmosphere cultivated around the pool. For most tourists, enjoying the pool is the closest thing

Figure 5.3   Doing 'nothing' by the pool

to the highly desired state of relaxation they are looking for. The mundanity of the pool has to be understood here as a negative quality that presupposes escape, relaxation and inversion. The pool is a relaxing environment that allows tourists 'to get away from the hustle and bustle', 'to switch off', 'to cut off', 'to slow down', 'to forget worries', 'to suspend the daily routine' and so on. However, the identification of the pool with relaxation is double edged. The fact that the pool is conceived as the perfect place to do nothing means it is also the most likely place to find boredom. What, at the beginning is rest and relaxation after four or five days becomes boredom and tedium. In the following passage, Linda describes very well the interdependence of relaxation and boredom. After six days the prospect of relaxing by the pool is no longer attractive. 'Well it's nice to sit but yes' – she admitted – 'now we are thinking oh another day by the pool! It's very nice but you can only drink so much, you can only eat so much. We've been around the same spas so many times. It's very nice but it's a little bit repetitive'. The way to combat boredom is to do something, which frequently involves hiring a car and travelling around. 'It was around Saturday when we decided to do something different. So really it's about a week ago when we thought we need to get away and hire a car. You need to do something else, if not it gets a bit tedious'. Tourists' comments paint a picture of the holiday as a succession of periods of excitement and boredom.

By the pool there is a thin line between relaxation and boredom. What initially is experienced as relaxation and rest soon becomes tedium and monotony. This is the case because there is no consistent utopia associated with the hotel pool, just a vague promise of luxury and hedonism. Such promises are ultimately nothing more than a temporal suspension of everyday rules and routines. The fact that the pool is quite empty of meta-narratives is nevertheless what makes the pool the perfect setting to find oneself. Hollowed out of utopias, obligations and constraints the pool offers the opportunity of inhabiting the mundane, making space for the cultivation of feeling, attitudes and ways of relating with others. The mundanity of the pool emerges here as a more positive quality relating to economies of affect. This means that boredom is not necessarily a sign of inauthenticity, it is also a sign of life. 'But what if one refuses to allow oneself to be chased away?', asks Kracauer – 'then boredom becomes the only proper occupation, since it provides a kind of guarantee that one is, so to speak, still in control of one's own existence. If one were never bored, one would presumably not really be present at all and would thus be merely one more object of boredom' (1995: 334). However the hotel – and here lays the paradox – does not like boredom, it is incompatible with the implications of people controlling their own existence. The hotel management makes sure that there is always an evening show, an organized excursion or a drinking game to draw the attention of the tourists and enliven the pool. As Kracauer reminds us, 'although one wants to do nothing, things are done to one: the world makes sure that one does not find oneself. And even if one perhaps isn't interested in it, the world itself is much too interested for one to find the peace and quiet necessary to be as thoroughly bored with the world as it ultimately deserves' (1995: 332). The hotel pool is simultaneously a palace of distraction and a site of boredom, a place to find the self and lose it again.

Conclusion

This chapter has examined the coastal hotel, one of the basic institutions of Mediterranean mass tourism. Rather than focusing on its structural logic, I have been concerned with the day to day uses of this space. My intention has been to produce a livelier account of the coastal hotel that sheds light on the everyday relations and practices through which this space is inhabited by tourists. In order to do so, I have moved away from ideas of commodification and alienation that emphasize the inhabitability of contemporary generic landscapes, considering instead the coastal hotel in terms of hospitality and affect. Derridian notions of hospitality (Derrida 2000) together with Maffesoli's ideas on affectual socialities (1996) open up mass tourism to new ways of analysis that are less hostile to this tourist phenomenon. The main appeal is the possibility of thinking the social life of the coastal hotel without recurring to notions of organic community. An ethnographic account of the pool – the main communal area of the coastal hotel – has illustrated the discussion. The pool has been presented as a space of conviviality, hedonism and relaxation the success of which depends upon the mobilization of mundane forms of hospitality and an ethics of conviviality. It is the main stage for the enactment of the warmly and welcoming atmosphere that identifies mass Mediterranean tourism. The social life of the pool – which consists mainly of fleeting mundane experiences orientated to the pursue of fun – have been revealed as more complex than expected. By the pool 'light touch forms of sociality' that emphasize informality and conviviality coexist with more enduring social forms that combine an ideological and an affectual dimension. By bringing together philosophies of hospitalities and hospitality studies, this chapter has brought to the fore a political dimension of mass tourism. Ultimately I have tried to demonstrate that the coastal hotel is an important site of politics, mainly related with the family and the everyday. The hotel is, paraphrasing Löfgren (1999), a cultural laboratory in which tourists are able to cultivate workable utopias, structures of feeling and new ways of living and relating with others.

References

Allen, J. 2000. On George Simmel: proximity, distance and movement, in *Thinking Space*, edited by M. Crang and N. Thrift. London: Routledge, 54–70.
Andrews, H. 2006. Consuming pleasures: package tourists in Mallorca, in *Tourism, Consumption and Representation: Narratives of Self and Place*, edited by K. Meetham, A. Andrews and S. Miles. Wallingford: CABI, 217–235.
Andrews, H. 2000. Consuming hospitality on holiday, in *In Search of Hospitality: Theoretical Perspectives and Debates*, edited by C. Lashley and A. Morrison. Oxford: Butterworth Heinemann, 235–54,
Augé, M. 1995. *Non-places: Introduction to an Anthropology of Supermodernity*. London: Verso.
Augé, M. 2000. Petrol stations, in *City A-Z*, edited by S. Pile and N. Thrift. London: Routledge, 177–9.
Bell, D. 2007a. The hospitable city: social relations in commercial spaces. *Progress in Human Geography*, 31(1), 7–22.
Bell, D. 2007b. Moments of hospitality, in *Mobilizing Hospitality: The Ethics of Social Relations in a Mobile World*, edited by J.G. Molz and S. Gibson. Aldershot: Ashgate, 29–44.
Binnie, J., Holloway, J., Millington, S., and Young, C. 2007. Mundane geographies: alienation, potentialities, and practice. *Environment and Planning A*, 39(3), 515–20
Buchanan, I. and Lambert, G. 2005. *Deleuze and Space*. Edinburgh: Edinburgh University Press.
Buhalis, D. 2000. Relationships in the distribution channel of tourism: conflicts between hoteliers and tour operators in the Mediterranean Region. *International Hospitality, Leisure and Tourism Administration Journal*, 1(1), 113–39.
Caputo, J.D. 1997. *Deconstruction in a Nutshell: A Conversation with Jacques Derrida*. New York: Fordham University Press.

Chambers, D. 2003. Family as place: family photograph albums and the domestication of public and private space, in *Picturing Place: Photography and the Geographical Imagination*, edited by J. Schwartz and J. Ryan. London: Tauris, 96–114.

Crang, M. 2002. Between places: producing hubs, flows, an networks. *Environment and Planning A*, 34, 569–74.

Derrida, J. 2000. *Of Hospitality*. Stanford: Stanford University Press.

Deutscher, P. 2005. *How to Read Derrida*. London: Granta Book

Goss, J. 1993. The 'magic of the mall': an analysis of form, function, and meaning in the contemporary retail built environment. *Annals of the Association of American Geographers*, 83(1), 18–47.

Gottdiener, M. 2001. *Life in the Air: Surviving the New Culture of Air Travel*. Oxford: Rowman and Littlefield.

Haldrup, M. and Larsen, J. 2003. The family gaze. *Tourist Studies*, 3(1), 23–45.

Harrison, P. 2002. The caesura: remarks on Wittgenstein's interruption of theory, or why practices elude explanation. *Geoforum*, 33(4), 487–503.

Hetherington, K. 1998. *Expressions of Identity: Space, Performance, Politics*. London: Sage.

Hirsch, M. 1999. *The Familial Gaze*. Hanover, NH: University Press of New England.

Hutchinson, S. 2000. Waiting for the bus. *Social Text*, 18(2), 107–120.

Ioannides, D. and Debbage, K. 1998. *The Economic Geography of the Tourist Industry*. London: Routledge.

Katz, M. 1999. The hotel Kracauer. *Differences: A Journal of Feminist Cultural Studies*, 11(2), 134–52.

Kracauer, S. 1995. *The Mass Ornament: Weimar Essays*. Cambridge, MA: Harvard University Press.

Latham, A. 2003. Urbanity, lifestyle and making sense of the new urban cultural economy: notes from Auckland, New Zealand. *Urban Studies*, 40(9), 1699–724.

Latham, A. and McCormack, D. 2004. Moving cities: rethinking the materialities of urban geographies. *Progress in Human Geography*, 28(6), 701–24.

Laurier, E. and Philo, C. 2004. Cafés and crowds, the problems of cosmopolitanism. [Online] Available at: http://web.geog.gla.ac.uk/olpapers/elaurier004.pdf [accessed: 4 May 2007].

Löfgren, O. 1999. *On Holiday: a History of Vacationing*. Berkley, CA: University of California Press.

Maffesoli, M. 1996. *The Time of the Tribes: The Decline of Individualism in Mass Society*. London: Sage.

Molotch, H. 2000. Pools (swimming), in *City A-Z*, edited by S. Pile and N. Thrift. London: Routledge, 189–91.

Molz, J.G. and Gibson, S. 2007. *Mobilizing Hospitality: The Ethics of Social Relations in a Mobile World*. Aldershot: Ashgate.

Mülder-Bach, I. 1997. Cinematic Ethnology: Siegfried Kracauer's 'The White Collar Masses'. *New Left Review*, 226, 41–56.

Reeh, H. 1998. Fragmentation, improvization, and urban quality: a heterotopian motif in Siegfried Kracauer, in *Intervals in the Philosophy of Architecture*, edited by A. Pérez-Gómez, and S. Parcell. London: McGill – Queen's University Press, 157–77.

Simmel, G. 1971. The stranger, in *George Simmel on Individuality and Social Forms*, edited by D. Levine. Chicago: University of Chicago Press, 143–49.

Simmel, G. 1997. The metropolis and the mental life, in *Simmel on Culture*, edited by D. Frisby and M. Featherstone. London: Sage, 174–85.

Sinclair, M.T. 1997. *Gender, Work and Tourism*. London: Routledge.

Tallack, D. 2002. 'Waiting, waiting': the hotel lobby, in the modern city, in *The Hieroglyphics of Space: Reading and Experiencing Modern Metropolis*, edited by N. Leach. London: Routledge, 139 – 151.

Thrift, N. 2004. Summoning life, in *Envisioning Human Geography*, edited by P. Cloke, P. Crang, and M. Goodwin. London: Arnold, 81–103.

Thrift, N. 2005. But malice aforethought: cities and the natural history of hatred. *Transactions of the Institute of British Geographers*, 30(2),133–50.

Urbain, J.D. 2003. *At the Beach*. Minneapolis: University of Minnesota Press.

Chapter 6
'De Veraneo en la Playa': Belonging and the Familiar in Mediterranean Mass Tourism

Javier Caletrío

Introduction

> Jacobo has been awake for some time but his eyes are closed. Light streams through the shutters into the stillness of his room, a stillness only broken by the occasional, intuitive movement of his leg reaching for the cool part of the bed. These are the sublime pleasures of summer. Lingering in bed, he slowly awakens to the familiar world around him: the smooth coolness of the untouched corners of the bed as the morning warms up, the redolent smell of old wood furniture, the sparrows in the garden and downstairs the voices of his grandparents talking over the first morning coffee. It is the first day of his holidays. He delays joining his grandparents to picture the summer stretched out before him, a summer of endless conversations on the beach, the promenade, the café, the garden. No more lessons or timetables, no more clocks or early nights. A summer to do nothing. Memories of summers past draw his lips into a faint smile. Finally he is on holiday, finally he is at home.

Jacobo will have breakfast with his family in the villa garden, help his parents with shopping in the local market and go to the beach at midday. But as he walks onto the beach his figure will not dilute into an impersonal mass of bronzed bodies that has become a paradigmatic image of Spanish Mediterranean beaches. Instead, Jacobo's walk along the shore will be a re-encounter with lifelong friends and acquaintances. His passage along the beach signals alternative historical-geographical itineraries in the evolution of mass tourism that do not fit well with established images of Mediterranean coastal resorts associated with two weeks package tours powerfully projected beyond popular discourse into the academic realm.

Jacobo is not a tourist, at least not in the academic sense of a person who leaves the familiar environments of everyday life in search of novelty or the exotic. Neither is he a *mass* tourist in the more everyday and dismissive sense of a passive member of a docile collective lured by the infamous trilogy of sand, sea and sun. Yet if people like Jacobo were to abandon mainland Spanish Mediterranean beaches, many would appear deserted and the image traditionally associated with mass tourism would fade away for these places are the domain not so much of package tourists as *veraneantes*, a Spanish expression that translates as 'those who regularly spend the summer here'.

Veraneantes's focus on re-encountering lifelong acquaintances and reproducing social networks problematizes a straightforward association of tourism with sightseeing and the consumption of place. Their largely domestic and often local origins, their ambivalent status as simultaneously hosts and guests, and their influence in setting the social tone in many resorts destabilize long held views of mass tourism in the Mediterranean as an invasive, corrupting force in what are portrayed as untouched, homogeneous and powerless local cultures. Looking at *veraneantes* helps to illustrate the complexity of Mediterranean leisure landscapes emerging from diverse, interrelated patterns of mooring and mobility, or dwelling-in-travelling (Clifford 1989, Urry 2000, Obrador 2003).

In the context of this collective volume this chapter aims at highlighting the significance and paradoxical effects of the familiar in the creation of Mediterranean landscapes. It should be noted, however, that while *veraneo* is a form of holidaymaking where certain forms of the familiar are expressed vividly, there is no intention to characterize *veraneo* in opposition to other travel styles where the familiar is present in a different, less evident or subtler guise.

There is no intention either to elaborate a new typology of tourist. Rather, my aim is to highlight a number of practices and discourses that produce particular places, affects and senses of belonging shared by diverse social groups, classes and nationalities. In so doing this chapter shows how *veraneo* gains expression through a number of local, national and transnational realities such as rural exodus, international migrations and colonial legacies. In this sense *veraneo* is a reminder of the need to look at tourism as a culturally and socially embedded phenomena whose uncertain boundaries are always on the move. *Veraneo* forcefully brings to the fore Adler's (1989) claim that the history of European tourism should be broadly seen as a history of myriad stylized configurations of sometimes new, sometimes old, social and cultural conventions, modes of travel, collective dreams and imaginaries, medical and scientific knowledge, hospitality and political institutions.

The data presented in this paper was gained in seven focus groups involving 41 participants and 30 in-depth interviews with *veraneantes* in four seaside resorts of the Costa Blanca, Spain, during the summers of 2000 and 2001. Interviewees ranged from social grades A (Status: upper middle class; Occupation: higher managerial, administrative or professional) to C2 (Status: skilled working class; Occupation: skilled manual workers) with most of the interviewees in Benidorm included in grades B (Status: middle class; Occupation: intermediate managerial, administrative or professional) and C1 (Status: lower middle class; Occupation: supervisory or clerical, junior managerial, administrative or professional) and interviewees in Altea, Calpe and Moraira in grades A and B.

The chapter is structured as follows. Firstly, I highlight the quantitative significance of the phenomenon of *veraneo* in Spain. Secondly, I briefly discuss how the familiar figures in social theories of travel and tourism. Thirdly, drawing on qualitative data from the Costa Blanca, I examine the role of *veraneo* in the encounter of Mediterranean tourism landscapes. There is a conclusion that includes suggestions for further research.

The Significance of *Veraneo*

Although information on the quantitative significance on *veraneantes* is scarce, data on second residences can provide a more reliable if indirect indication of the massive scale of the phenomenon. Of a total of 11 million households in Spain, 21 percent have access to a second residence, owned or leased by relatives and friends (IET 2000). During 2000, Spanish residents made a total of 123 million overnight trips equalling a total of 562 million nights. 62 percent of these overnight journeys were short trips of less than four nights to a second residence.

International demand of this type of accommodation is also notable. In Majorca the percentage of visitors using a second residence over the total arriving at the island is 67 percent for Germans, 27 percent for British, 66.5 percent for French, and 44 percent for Spaniards (CAIB, 1997; Duhamel, 2000). An indication of the degree of familiarity with the place visited is given by the rates of ownership. More

Table 6.1   Rate of ownership of second residence and number of beds in relation with hotel beds

Resort	Number of second residences	Beds in second residences	Hotel beds	Ownership second residences
Oropesa	7,600	30,360	1,849	75%
Cullera	14,264	57,058	1,026	64.9%
Benidorm	38,333	153,332	33,652	27%

Note: In showing a relatively high rate of ownership and an overwhelming predominance of accommodation in second residences, the resorts of Cullera and Oropesa are representative of most seaside resorts in the region of Valencia. When taking the rate of ownership as an indication of the degree of familiarity with the resort it should be noted that those renting an apartment or a villa frequently are habitual visitors too, though usually with less financial possibilities to acquire a second home. Benidorm represents the exception to the largely residential model of tourism development predominant in the Valencia region. Its 33,652 hotel beds in 125 hotels represent just over forty percent of the total hotel capacity. Though lesser than in the other two resorts included in the sample, the rate of ownership of second residences is still significant. Users of rented apartments include habitual visitors (families frequently book the same apartment year after year) and package holidaymakers (mainly British) making use of self-catering accommodation.

Source: Agencia Valenciana del Turisme, 1999.

than 70 percent of north Europeans making use of this type of accommodation in Spain were using their own or a friend's or relative's while the remaining 30 percent were renting it (IET 2000). The Institute of Foreign Property Owners and associations of building promoters estimate that in 1998 the total number of Europeans with a property in Spain was 1,100,000. That figure increased to 1.5 million by 2001 and was estimated to reach 1,750,000 by 2003 (FIPE 2002).

The phenomenon of second residences in the region of Valencia where field work for the research presented here was undertaken is particularly prominent. According to reliable data for 1992 provided by the Secretaría General de Turismo, in the region of Valencia there were 78,210 hotel beds and 2,219,000 non-hotel beds (self-catering, second homes and unregistered supply) (Monfort and Ivars 2001: 28). These figures include both inland and coastal areas. Another study gives a maximum figure of 1,493,705 beds for tourist use along the main coastal resorts, 92 percent of which correspond to non-hotel accommodation (for example villas, apartments and terraced houses) (Fundación Cavanilles 1994). In Valencian seaside resorts most second residences are owned rather than rented with the exception of Benidorm (AVT 2000b). Reliable comprehensive data on rates of ownership is lacking though the partial research available for two highly representative resorts is 75 percent and 64.9 percent (AVT 1999).

This predominance of *veraneantes* in many resorts remains largely unacknowledged by scholars writing on tourism in the Mediterranean. For example, despite his perceptive chapter on summer vacations and second homes – what he calls 'cottage cultures' – in Scandinavia, Löfgren's (1999) fails to acknowledge its popularity in southern Europe and goes on to characterize the contemporary Mediterranean as a region that lives to the rhythm of the two-week package tour. Yet the practice of *veraneo* confers coastal landscapes in the Spanish Mediterranean with a different rhythm. Or to be faithful to the Braudelian analogy used by Löfgren, Mediterranean life palpitates not with one but a myriad rhythms, some of which stem from, revolve around or become entangled with the practice of *veraneo*.

The Lure of the Familiar and Tourism Studies

The rhythms of veraneo take shape and meaning in relation to other places and forms of leisure and work. 'Altea', a veraneante tells me, 'is not a place to do tourism, or seeing things, it is a place to stay, to rest, to meet people'. In its blunt simplicity the statement hints at a relational sense of place and the three key discourses that structure veraneo as a travel style: relaxation ('being lazy'), socializing ('we're a family'), and attachment to place ('this is home'). A standard day of a veraneante passes by socializing with relatives and friends in a relaxed way (on 'laid-back mobilities', see Haldrup 2005). And that life involves spatially restricted patterns of mobility that gravitate around the beach and the promenade, the scenario of a range of usually quite mundane practices such as walking, eating and drinking in bars and cafeterias or sun bathing and swimming in the sea, repeated with surgical precision day after day from the beginning until the end of her holidays:

> The house is somehow superfluous to us. You go to the beach in the morning, meet your friends, then you may have a beer together and swim for a while and then have a salad, and then sometimes the conversation with friends is good and you decide to eat some sardines here in the bar instead of eating at home [the villa]. So you may stay the whole day here. And then at night you may ask Angel, the owner of the bar, to cook a paella for us all and afterwards you may have a dip in the sea and you may return home at one or two in the morning. That's how my summer goes by (Jesús, 45, from Valencia, Spain, bank manager, regular visitor to Moraira).

As the quotes illustrate, daily life is punctuated by highly valued quotidian rituals such as the three main daily meals, siesta, having a drink in the evening in the promenade and engaging in conversations until late at night with friends and relatives, all of which convey a sense of the familiar. It is precisely those apparently banal and highly predictable rhythms and patterns of daily activity in the company of significant others what give special meaning and value to their holidays and place of vacations.

In trying to make sense of this predictability and familiarity through social theories of tourism one is led either to discard *veraneo* as a form of tourism or dismiss it as a failure to accomplish its emancipatory potential. Issues of familiarity have indeed been largely overlooked in tourism literature. In assuming a home and away dichotomy whereby home is associated with the familiar and away with novelty and the exotic, the familiar has tended to be understood as what the tourist leaves behind. In one of his earlier writings on the topic, Urry (1990: 2–3) contends that tourism involves movement of people to 'new place or places', that tourist sites are 'out of the ordinary' and that there is 'a clear intention to return 'home'. When the familiar appears in the tourist experience travel literature tends to see it as the result of the increasing standardization of everyday life. Accounts of tourism as a quest of authenticity associate the familiar with commodification and a loss of authenticity (MacCannell 1973). This line of argument is furthered by Ritzer in his McDonaldization thesis whereby the increasing commodification of tourist destinations results in or is expressed through highly standardized, familiar, predictable and controlled tourist products, services and experiences (Ritzer 1996). Turning the focus on the tourist, authors such as Bruner (1991) or Rojek (1994) raise doubts about the potential of tourism for self-development or as a 'way of escape' for tourists inexorably carry with them a cultural 'baggage' as members of a specific class, gender, race, or nation.

An attentive reading of histories of travel and tourism suggests, however, multiple and complex entanglements between home and away (see Pemble 1987). This entanglements and the ontological work that cultural baggage does on places is increasingly being acknowledged through the performative turn in tourism studies (Edensor 1997, Coleman and Crang 2002, Bærenholdt, Haldrup, Larsen, Urry 2005). In an attempt to rethink the place of travel in modernity, Minca and Oakes argue that 'the study of travel must also raise questions about the meaning of home, about belonging, about how places get made and remade' (2006: 1). They see place as integral to travel and travel as integral to place and home-making practices: 'place-making is fundamentally infused with travel and all the baggage that gets shipped along the way – difference, strangeness, alienation, nostalgia, homesickness, inspiration, fear, frustration, hopes, and expectations fulfilled and dashed' (2006: 1). In a similar vein, feminist and postcolonial critics call for an alternative 'framework for rethinking home and migration in ways that open out discussion beyond oppositions such as stasis versus transformation, or presence versus absence' (Ahmed et al. 2003: 1). Elaborating this argument in relation with tourism, Crang notes that 'the tourist seeks to travel to be present at a place, but as we examine those places we find they are shot through by absences where distant others, removed in space and time, haunt the sites' (2006: 49). Such attentiveness to the complex and contingent relation between home and away also shows how for many people travel is also about *getting home* when the aim of the journey is visiting family and friends scattered across different places (Larsen et al. 2007, Urry 2000). It is this path of disrupted categories and dualisms that I will follow to examine the familiar in the context of *veraneo*.

Performing the familiar: *habitus* and the beach

The practice of *veraneo* is paradigmatic of forms of travel styles where destination is home and meeting significant others is crucial. So entangled people and place appear in *veraneantes* accounts in the Costa Blanca that one term often stands for the other:

> Bea: Here the beach is very good, and calm and tranquil and the nightlife on the other side of the beach is not bad too. We like Benidorm a lot. [...]
> Andrea: You know what happens? What happens is that we also have a very strong affective attachment to Benidorm.
> Pierre: No, *also* is not the right word. The word is *above all*.
> Bea: Yes, you are right.
> Alfredo: It is a whole set of things. There are people that you really appreciate and whom you've known for a long time. These are people that you can only see once a year and you can only see them here.And beside this the tranquillity ....
> But yes, the feelings towards the place are very important. You have to bear in mind that we have been coming here for many years. [...] Of course we like the beach but when you ask for the reasons why we come here, we like the beach, the sun, everything, but for us, well at least for me, it is my people.
> Bea: What we come for is this beach, this sun, this breeze...with these people.
> Pierre: We cannot conceive any of those things separated from the rest.

Alfredo: When Andrea says "I need Benidorm", she's not referring to Benidorm itself or just this beach in particular. She means that she needs to come to Benidorm, to this place, this beach, and see this person and the other person.
(Young professionals from Madrid, Navarra, Spain, and Paris, France, regular summer visitors to Benidorm, Playa de Poniente).

These young professionals show a reluctance to prioritize place over people and present both as inextricably entangled. This kind of narrative was often accompanied by the clarification that seeing their friends in their place of residence in the winter months 'felt different' and lacked the excitement it had in the summer. The communion of place and people underpins home-making practices and veraneantes's sense of place in the Costa Blanca. But such relationship is contingent. The obvious but important point is that, inevitably, people change and places change too. This is something that Löfgren (1999) fails to sufficiently acknowledge in his otherwise excellent description of 'cottage cultures' in Scandinavia and the USA. Since a veraneante's engagement with a resort is a long term, often lifelong affair, negotiating that relationship through one's biography and through generations in a family is an essential part of the experience of veraneo. The rest of the chapter offers a necessarily brief introduction to the dynamics of this relationship in the Costa Blanca. An entry point to examine this issue is the way the beach is encountered as a familiar place. The beach is the public space par excellence in seaside resorts and the place where that 'affective attachment' Andrea refers to is expressed most vividly.[1] In my research the beach is also the place where most interviews with veraneantes took place.

The beach is a familiar place in the double sense of a well known place and a family place, a place where *veraneantes* socialize with their families and where families interact with other families. The idea of the beach as a familiar place is continually recreated through the practices that *veraneantes* perform during their holidays and they usually go about their daily lives without consciously thinking about it, it is experienced as a routine. Yet such a specific way of encountering the beach has to be learned (Lenček and Bosker 1998, Löfgren 1994). It implies acquiring competence in a number of activities and ways of seeing, smelling, touching, listening and moving around deemed to be appropriate, *natural*, on that beach: visiting the beach every morning at around the same time, walking along the shore in a certain manner, placing the beach towel and the parasol in the same place every day, looking at people and the sea in specific ways, talking with friends and relatives, re-encountering them every year or every few weeks, sharing memories – last night's party or childhood anecdotes – making children

---

1   While spaces with restricted access such as elite tennis and golf clubs and marinas are an emergent trend, the beach and the promenade are still the places for socialising for the majority of veraneantes, especially in resorts where apartments are the main accommodation type such as Benidorm, Cullera, Gandia, and Oropesa. The beach is the public place where they spend most of their time and often one of the few open public spaces.

and youngsters participate in such story-telling, swimming to particular places, collecting specific plants or animals, etc. Through ongoing engagement with that set of practices, *veraneantes* acquire a disposition to act, sense and feel the beach in a specific way.

Such disposition to live the beach as familiar is something they acquire as a *habitus* and is expressed as an easiness at performing tasks and relating with place and people. In the Costa Blanca affinity in the economic and cultural capital displayed by tourists and retirement migrants is important in determining social relationships and in this sense *habitus* needs to be understood in Bourdieu's (1984) terms as the practical and affective dispositions shared by a social class and acquired in the fight for resources in a field. The Costa Blanca is punctuated by a series of seaside resorts displaying their own specific social ambience and characterized by struggles over distinction. When talking about their place of holidays, *veraneantes* invoke a relational frame of awareness about other resorts, often outside Spain: 'What we want is tranquillity to feel the time passing by. That's something you can feel in Altea but not in the Côte d'Azur, the Côte d'Azur is like Paris'. Resorts are classified in scales of taste and naturalness where Benidorm invariably ranks in the lowest position: 'Altea is not like Benidorm, this is a family place, more natural' or 'obviously Altea is different from Benidorm'. Benidorm figures as the 'chaotic', 'massified', 'tacky' place where 'you can smell the beach in the summer, you smell the suntan oil', 'the status of the people is so low', 'people hang their laundry in the windows', a place in sun, which 'honestly, I really think is unbearable'. *Veraneantes* in Playa de Poniente in Benidorm eagerly stress that *their* beach is not like Playa de Levante where the 'British hooligans' or simply 'the tourists' stay. In *veraneantes*'s accounts 'the tourist' and 'the masses' are always to be found beyond the boundaries of one's place although their threat is never far enough and always seems imminent.

Scholars relying on Bourdieu's theory of the judgement of taste portray resorts as sites for performing identities and assume intense competition between places to attract increasingly mobile and discerning tourists, a result of which is an increasing volatility of the fortunes of tourist destinations (Urry 1990, Munt 1994, Shaw and Williams 1997). Yet *veaneantes* show a stunning loyalty to place, even after the loss of status due to perceived excessive building. This suggests more complex dynamics at play in the encounter of place. When Eva describes Altea as a place that 'is not like Benidorm' and adds that 'it is a cute little village, and besides that *you feel at home*' she is participating in a classificatory struggle seeking to distinguish herself from 'Benidorm tourists', 'vulgar tourists' or just 'less distinguished' visitors. But she is also expressing something more complex, personal and specific that goes beyond or before or along with class politics and conventions. Eva's sense of easiness or comfort partly stems from the ongoing engagement with the materiality of place and the embodied activities that it affords. This affective disposition has more affinity to Mauss's (1979) than Bourdieu's habitus. Clearly Mauss's habitus or *techniques of the body* are intrinsically social but they are less determined by class and leave more space for innovation and change and a more comprehensive array of contexts and experiences shaping bodily and affective dispositions. As a result, the sense of distinction derived from a *veraneante*'s competence in the use of the beach does not necessarily involve elitism or snobbishness, at least not always or not in every context and situation. Ana's experience of the beach points in this direction:

> People of my age have had a very close relationship with the beach and the sea. We knew every centimetre of the beach, we knew the rocks over here, the rocks over there, the seaweed prairies and so on. We knew it extremely well. We had a very deep knowledge of it. [...] We would go to the other part of the beach where the coral was, we knew that in that other place we could find the other type of shells that we call *orejitas*, we knew the site where the limpets dwelt, the site where the sea urchin dwelt. It was as if, you know, when you get into your own house where you know where your TV is, or any other stuff is, do you know what I mean? We knew it very well. The sea has been a very important part of our lives here, a very important part. I could tell you [...] lots of stories around the sea, many, many, many (Ana, 51, from Valencia, Spain, regular visitor to Moraira).

Belonging and Nostalgia

Ana's analogy of the beach with her home conveys a sense of familiarity and intimacy. It is a sense of familiarity stemming from practical engagement with the materiality of the beach in company of others. She is referring to a kind of experience in which 'the landscape is with us, not against us, just as we are part of it' (Ingold 1993: 154). This familiarity with the landscape involves an attachment to place which is often captured by the expression *mi pueblo*. The term *pueblo* literally means village. *Mi pueblo* translates as my hometown or my family's hometown and evokes a sense of belonging and a place where one feels welcomed by familiar faces:

> After coming to Benidorm for so long you end up seeing it as other urban people see or feel their parents's pueblo, a place that you really appreciate. My parents don't have a *pueblo* because they were born in Lleida, so Benidorm is my *pueblo*. (Sonsoles, 24, from Lleida, Spain, regular summer visitor to Playa de Poniente since she was born)

> I feel very attached to this place. This is the place that comes to my mind when I ask myself: "where will I retire?" I say: "I'll come to Moraira." [...] I see Moraira as if it was my *pueblo*. Every time I come back to Moraira I feel as if I was coming back to my *pueblo*. Do you know what I mean? Do you know about the people who had to migrate from their pueblo when they are young and return every year for holidays to meet their family, to reencounter their place? [...]

> Every time I come here it is as if I was coming to my *pueblo*. I have to work in
> Valencia but every weekend I come to my *pueblo*, to the place where … to my
> roots (Ana, 51, from Valencia, Spain, regular visitor to Moraira).

The affective content of the words *mi pueblo* is related, as Cristina points out, to
the rural exodus that has drained large parts of rural Spain of their population,
a process that reached massive proportions during the 1960s and 1970s and is
still a cause of concern in many areas. This demographic decline has occurred
all across rural Europe with the more agricultural southern countries amongst
the most severely affected. As a result, significant parts of urban populations in
Spain, Italy and Greece have now access to a second residence in the countryside,
belonging either to them or their relatives. Home towns and rural areas are a
persistent presence in the life of many urban families, not only through frequent
visits to relatives and short leisure breaks, but also on a more daily basis
through, for instance, consumption of local, frequently hand made, foods and
drinks. Everyday allusions to home villages are not uncommonly suffused with
nostalgia and travelling at least once a year, particularly at the time of the local
summer festival, is felt as an obligation and a way of 'not losing touch with one's
roots'.

For those born in a village, the expression *mi pueblo* is both a statement
of a fact – a identity acquired by chance – and an affective attachment – an
active choice. In coastal resorts the majority of *veraneantes* are just claiming an
affective attachment and a willingness to belong. But often it is a willingness to
belong determined by earlier family choices. In understanding Ana's investment
of Moraira with a sense of home, the focus needs to be placed not so much on
her but on the generational dynamics of her family. The practice of *veraneo* is
a family business whose choreography includes differences in lived rhythms
according to age and gender and particular patterns of activities and attachment
to (or detachment of) family life and place. This connection with family life
is crucial to grasp the temporal logic of *veraneo* and leisure landscapes in the
Spanish Mediterranean.

To illustrate this point I turn to Beatrice, a French citizen who has visited
Moraira for more than thirty years. Since Beatrice and her husband retired in the
late 1990s they have spent the spring and summer in Moraira, and the autumn
and winter in Pau near the Pyrenees skiing and walking, something she considers
an essential part of her winter life. Now she describes herself as 'a person with a
double life, one in France and another one in Moraira'. In the following excerpt
she describes this sense of belonging as involving growing roots in different places
in a rhythomatic manner:

> Beatrice: I knew that when children grow what they like is not being constantly
> travelling to different places but seeing always the same friends, do sports, you
> know, to have roots, another roots. What they like is changing to a different
> place but still having roots. For instance, I don't like that habit of many French

people who visit a different place every year. I rather like to have my own house. I rather like to change but without changing. And we still travel to other places but at another time of the year, not during the summer.

JC: Changing without changing?

Beatrice: Yes, we change place, we change everything, we change the language, food, I also change my readings because I read a lot in Spanish, and we change friends. And yet we still come to our house, to a place where we have roots. (Beatrice, 60, retired, from Pau, France, regular visitor to Moraira).

To the extent that the aesthetic qualities and social tone of the place influenced Beatrice's decision to acquire a property and 'grow roots' in Moraira, hers can be seen as a form of 'elective belonging' as conceptualized by Savage et al. (2004) whereby attachment to place is not derived from a familiarity with a face-to-face community but from a relational sense of place, a capability to assess a place in relation to others and fit one's biography within its social, economic and cultural dynamics.

For Beatrice affordances of place are key in her relational frame of awareness, especially the affordances of the beach: 'at that moment I had two adventurous boys and I thought Las Landas [French resort where she owned a second residence] was not a safe place for us to spend our holidays'. As her sons were growing older her time in Moraira, where the lack of sea currents makes swimming safer, grew longer. Her sons preferred Moraira as a holiday place for the freedom they enjoyed there and finally she sold the house in Las Landas. Parenting is central in the performances of *veraneo* and in this case the affordances of place are important in fostering a sense of attachment (see Savage et al. 2004: 62 on belonging, parenting and the education field). Interviewees in other resorts often highlighted how the 'familiar' character of the beach and its 'gentle' physical attributes provided tranquillity and relaxation to parents who did not need to watch their children closely.

Its sensitivity to the way in which people make sense of place in relation to their biographies confers the concept of elective belonging with temporal dynamism. Yet the experiences of *veraneo* demand a greater attention to generational change and emergent social networks in generating a sense of belonging. After 30 years visiting Moraira and thanks to her and her sons' ability to speak Spanish, Beatrice's social networks include 'natives' – born and bred in Moraira – numerous Spanish and non-Spanish *veraneantes*, and retirement migrants. One of her sons married a Spanish *veraneante* and they too visit the resort regularly and maintain their social networks. Social relations in Moraira are more extensive and resilient than the ones left behind in Las Landas. Moraira often becomes a family meeting place as her sons now live in different locations in France. She has also learnt to accept change in the aesthetic qualities and social tone of the place and fidelity to a face-to-face community complements and on occasions compensates for the loss of such qualities.

There is another aspect of Beatrice's decision to 'grow roots' in Moraira that cannot be missed. She lived in Algeria for four years and retains special affection for its landscape which has climatic, botanical, geological and cultural affinities with the Costa Blanca. The affordances of place are significant in bringing alive her memories in the North of Africa. Such mnemonic potential of the Mediterranean landscape is explicitly referred to in the following excerpt where Marie and Alice, *pied noirs* – north African-born French who migrated to France after the colony's independence – explain what led them to decide to 'grow roots' in Altea: 'We bought the apartment here to grow roots. For thirty years we've been coming here with an inexhaustible joy and pleasure. [...] We have the feeling that this village belongs to us'. Later on in the interview, when talking about their feelings towards the sea they elaborate the significance of spending their holidays by the Mediterranean Sea:

> Alice: We were born by the sea and now we live in Paris [...].We have lived by the sea in the Maghreb, Casablanca, Tangier, then Marseille and then Paris which is another sea populated by people. But I cannot live without Paris. [...] To me the sea means freedom, the colour.
>
> Marie: And the quality of life, the different lifestyle of the Mediterranean. I mean that it reminds us of our roots. You know, in Algeria and Morocco there were also many Spaniards and people from other places and we carry that mixture with us, we are a kind of synthesis of that mixture.
>
> Alice: The Sea has fed our youth, our adolescence, our childhood. You know, the memories of your childhood are very important. When we see this sea we feel good, we revive something of our adolescence, perhaps subconsciously.
>
> JC: Do you mean that you feel that when you see this sea or any sea in general?
>
> Marie: Look, the Atlantic is very nice, but it rains and the water is cold and it's very different. I'm talking about the colours of this sea, its light, and the perfumes of the countryside here, the orange blossom, the jasmine. When you walk along the beach of La Olla you pass next to an orange grove and a very rich aroma comes to you. That smell is not in the Atlantic. (Marie and Alice, forties, from Paris, regular summer visitors to Altea for 35 years).

The significance of the Mediterranean Sea and landscape for Marie and Alice partly lies in eliciting reminiscences of their childhood in North Africa. This is afforded by a sensuous encounter with the landscape. The aroma of the orange grove brings Marie childhood memories and a sense of place. We may understand this as an example of what Hetherington calls the *praesentia* of place (Hetherington 2003, Obrador 2007). *Praesentia* is concerned with presence and confirmation. In this case the aroma of the orange blossom is not standing as a metaphor for her childhood but affording the presence of something absent – her childhood memories. Through the otherness of the materiality of the orange grove and its aroma, a memory is made present. Through the feel of the aroma a *praesentia* is revealed. What is revealed is a memory but also a sense of the familiar, a feeling of something or somewhere known. Marie notes the aroma of the orange grove but she also talks about other senses: touching a warm sea, appreciating the texture of colours and light.

In her ethnography of *pied noirs* travelling to Malta, Andrea Smith (2006) describes similar reactions to the landscape which is so reminiscent of North Africa. The prickly pear in particular is the plant that most lively reactions provoke:

> This revelling of nostalgia for the North African landscape culminated on the last day of the trip. While waiting for the ferry to take us from Gozo to Malta, we parked on a hill overlooking the bay. It was cold and damp, and the sun was setting. But just next to the parked vans along the side of the road one of the most abundant stands of Barbary figs we had seen. It was just too much. Giggling like children, the elderly settlers in the van ahead of me began making a commotion, taking photographs of each other and rummaging through their bags. It was time for a Barbary fig feast! [...] Everyone began gorging themselves on the delicious ripe fruits [...]. One rather tense, brittle woman began to talk with enthusiasm for the first time on the trip. "Oh, this is just like my youth!" she kept exclaiming. "You known, it has been thirty-three years since I've eaten these", she told me. "Thirty-three years. How much they remind me of the good old days!" (Smith 2006: 218–9).

Such a Proustian moment is further elaborated by Giuseppe, a migrant from Italy who settled in Altea after coming as a tourist and now spends his holidays in a second home in his hometown in Calabria. In the excerpt below he talks effusively about sensuous experiences: the taste of certain vegetables he picks on the shore or the raw mussels he eats with drops of olive oil and lemon juice, the touch of pebbles, or the slime of the octopus. Through this performative repertoire Giuseppe is not 'making sense' of the landscape. Rather, in his encounter with its materiality he is gaining a sense of who he is and where, he is recognizing himself as a particular subject. Through this sensuous performative repertoire Giuseppe is gaining a sense of place. He elaborates this point when asked to talk about the sea:

> What the sea brings to me ... this question is related to what I was telling you earlier about the fact of identifying yourself with the Mediterranean. [...] What it brings to me is so many things, so many things I have assimilated since early childhood. My first memory of the sea, that's something I'll never forget. The smell of the sea ... There are so many things ... for instance things related with food, I have spent so many nights by the sea eating certain food: tomatoes, water melons, olives, grapes. All those things are very related, aren't they? The smells, the wild flowers, the wild vegetables that you pick on the rocks of the shore. Where I was born in the south of Italy this wild vegetable is picked during the summer and kept in vinegar. It is a type of seaweed that grows next to the water ... this type of vegetable I used to eat with a certain kind of food in certain

moments and therefore you cannot but feel strongly identified with it. It happens the same with fried little fish, with the octopuses. For instance I'm referring to the feeling of capturing an octopus and smashing it thirty or forty times on a rock on the shore in order to soften its flesh. And by doing that it produces slime, a kind of surf that is sticky, a sticky secretion and you get that when you clean the octopus. What I'm trying to tell you is that all those things are sensations that by being and growing next to the sea you can live, sensations that gradually become part of you and you enjoy to the core, through and through.

The taste of the mussels, the grapefruit, the melon, the olive oil, the smell of the wild flowers, the sea, the touch of the octopus, all these sensuous experiences create an experience of the beach as a place where landscape and the body come together. This experience of landscape is specific, local and partial and is concerned with confirmation rather than representation. This proximal form of knowing, Hetherington argues, 'helps to perform a located subjectivity constituted as a partial contact and involvement in the materiality of this Other' (Hetherington 2003). Such embodied experiences of the beach as a familiar place are not available to anyone. It requires certain ability in the use of proximal and performative forms of knowledge. This ability is acquired as a matter of routine and incorporated as a *habitus*.

Conclusion

Along with the liminal landscapes of Mediterranean beaches playfully encountered by many tourists (Inglis 2000), there are other overlapping or intersecting landscapes that revolve around a reencounter of place and people, landscapes of memories, heavy with time, which speak of the work invested by holidaymakers and migrants in creating a sense of home. Home is an essential part of travel and this applies not just to the practice of *veraneo* but other mobilities traversing and making up the Costa Blanca: the Moroccan or Senegalese migrant in Benitatxell longing for *papers* to begin a life in Europe and thinking about relatives left behind, an Algerian family resting at the petrol station in Moraira in their annual journey to the Kabila to re-encounter friends and relatives, a Parisian family making the best of a week in Altea to heal family wounds, British retirement migrants recreating a *little England* or a cosmopolitan Mediterranean. In bringing with them stylized forms of feeling, sensing and remembering nurtured in other geographies and times, tourists and migrants turn Spanish landscapes into contingent, multifaceted realities.

The accounts presented in this chapter suggest that *veraneantes*'s encounter of place in the Costa Blanca is not a simple, linear process, partly because coastal resorts are lived as complex, paradoxical locales. Rather than being experienced as spatially fixed containers of social processes, resorts are being lived as entangled in a complex relationality of other places and landscapes. Resorts, we might say, are not fixed locales but topologically complex entities put on the move in a variety of ways. Part of this chapter has highlighted bodily and material ways in which geographies and landscapes are mobilized. It has described how through sensuous dispositions and acts of remembrance remote geographies and/or times are performed as 'absent presences' that become interwoven within tourist resorts. The heterogeneity of these crumpled time-spaces, the increasingly warped and twisted topologies associated with and enabled by these pattern of dwelling-in-travelling, render the experience of landscape in seaside resorts a complex and multidirectional process.

The role of memory and social networks helps to explain why, with the alleged shift from modern to postmodern sensibilities, Mediterranean seaside resorts continue to exert such an inexhaustible appeal, why after so much apparent environmental destruction and the loss of status of overbuilt resorts, the *Mediterranean passion*, to borrow Pemble's words, is still alive for working, middle and upper classes. Obviously this is not the only reason in explaining the appeal of resorts and the power of distinctions of taste and economic and political processes should not be undervalued. What is needed is more empirical research on the relationship between class related judgements of taste, affordances of place, memory, social networks, and family change at an individual and societal level. This would allow articulating more dynamic notions of elective belonging.

With the boom in property development in the late 1990s seaside resorts have become more internationalized with around 50 percent of the new state development owned by non-Spanish EU citizens, and a significant proportion of the labour provided by North African, East European and South American migrants. The extent to which this is leading to the internationalization of social networks which are so central to the practice of *veraneo* is not clear, though evidence suggest that these mobilities largely bypass each other rather than intersect or co-amplify their emancipatory potential. With its recent eruption in the social sciences, the concept of cosmopolitanism is opening potentially fruitful lines of enquiry to explore this question (see Beck and Sznaider 2006). Yet, whether growing patterns of mobility in the Costa Blanca are creating a potential to realize more progressive societies or generating greater awareness about the transnational interdependencies remains to be seen.

The greatest difficulty in answering this question is the lack of historical evidence about the intensity and qualities of cosmopolitan dispositions over time and how these are enabled by social networks and encounters of place. For all the recent emphasis about 'history matters' and despite the cultural turn in tourism studies and its emphasis on the historical nature of travel styles (Adler 1989), research in tourism remains surprisingly ahistorical. In this sense what is needed are not so much histories of tourism but a *historical* tourism studies that is sensitive to the *longue durée* of demographic, socio-environmental, technological, cultural, economic and political processes underpinning travel and tourism (on historical social sciences, see Wallerstein 2000).

References

Adler, J. 1989. Travel as performed art. *American Journal of Sociology*, 94(6), 1366–91.

Agència Valenciana de Turisme, 1999. *Aproximación a la Oferta y Demanda de Viviendas de Uso Turístico en los Muncipios de Benidorm, Cullera y Oropesa.* Valencia: AVT.

Agència Valenciana de Turisme, 2001. *El Turismo en la Comunidad Valenciana 2000.* Valencia: AVT.

Agència Valenciana de Turismo, 2002. *El Turismo en la Comunidad Valenciana 2001.* Valencia: AVT.

Ahmed, S., Castañeda, C., Fortier, A-M., Sheller, M. 2003. *Uprootings/ Regroundings.* Oxford: Berg.

Beck, U. and Sznaider, N. 2006. Unpacking cosmopolitanism for the social sciences: a research agenda. *The British Journal of Sociology*, 57(1),1–23.

Bourdieu, P. 1979. *Distinction.* London: Routledge.

Bruner, E. 1991. Transformation of self in tourism. *Annals of Tourism Research*, 18(2), 238–50.

CAIB 1997. *Despesa Turística*, Mallorca: CAIB.

Crang, M. 2006. Circulation and emplacement: the hollowed-out performance of tourism, in *Travels in Paradox*, edited by C. Minca and T. Oakes. Oxford: Rowman and Littlefield.

FIPE. 2000. Turismo residencial en España. *Boletín FIPE*, 5–6.

Inglis, F. 2000. *The Delicious History of the Holiday.* London: Routledge.

Hetherington, K. 2003. Spatial textures: place, touch, and praesentia. *Environment and Planning A*, 35(11), 1933–1944.

Ingold, T. 1993. The temporality of landscape. *World Archaeology*, 25(2), 152–74.

Instituto de Estudios Turísticos. 2000. *Encuesta Turística.* Madrid: IET.

Kaplan, C. 1997. *Questions of Travel: Postmodern Discourses of Displacement.* London: Duke University Press.

Lenček, L. and Bosker, G. 1998. *The Beach.* London: Pimlico.

Löfgren, O. 1999. *On Holiday: A History of Vacationing.* Berkeley: University of California Press.

Löfgren, O. 1994. Learning to be a tourist, *Ethnologia Scandinavica*, 24, 102–5.

MacCannell, D. 1999. *The Tourist.* New York: Schocken.

Mauss, M. 1979. *Sociology and Psychology: Essays.* London: Routledge.

Monfort, M. and Ivars, J. 2001. Towards a sustained competitiveness of Spanish tourism, in *Mediterranean Tourism: Facets of Socioeconomic Development and Cultural Change*, edited by Y. Apostopoulos, P. Loukissas and L. Leontidou. London: Routledge.

Munt, I. 1994. The 'Other' postmodern tourism: culture, travel and the new middle classes. *Theory, Culture & Society*, 11(3), 101–23.

Obrador Pons, P. 2003. Being on holiday: tourist dwelling, bodies and place. *Tourist Studies*, 3(1), 47–66.

Obrador Pons, P. 2007. A haptic geography of the beach: naked bodies, vision and touch. *Social and Cultural Geography*, 8(1), 123–41.

Pemble, J. 1987. *The Mediterranean Passion*. Oxford: Clarendon.

Ritzer, G. 1993. *The McDonaldization of Society*. London: Sage.

Ritzer, G. and Liska, A. 1997. 'McDisneyization' and 'post-tourism': complementary perspectives on contemporary tourism, in *Touring Cultures*, edited by C. Rojek and J. Urry. London: Routledge.

Rojek, C. 1993. *Ways of Scape*. London: Routledge.

Savage, M., Bagnall, G. and Longhurst, B. 2004. *Globalization and Belonging*. London: Sage.

Shaw, G. and Williams, A. 1997. *The Rise and Fall of British Seaside Resorts*. London: Thomson Learning.

Smith, A.L. 2006. *Colonial Memory and Postcolonial Europe: Maltese Settlers in Algeria and France*. Bloomington Indiana: University Press.

Urry, J. 1990. *The Tourist Gaze*. London: Sage.

Wallerstein, I. 2000. From sociology to historical social science: prospects and obstacles. *The British Journal of Sociology*, 51(1), 25–36

Chapter 7

Hosts and Guests, Guests and Hosts: British Residential Tourism in the Costa del Sol[1]

Karen O'Reilly

When Pau Obrador first asked if I would like to contribute to this book on tourism cultures in the Mediterranean I immediately said yes. I have been studying and writing about British migrants in Spain for over ten years, and have been a member of the community as a second-home owner or peripatetic migrant for much of that time.[2] Most of my work has been at pains to point out that these migrants do not see themselves as tourists. They mark themselves out in opposition to tourists. They say 'we live here' or 'we are here to stay'. They talk about tourists as 'other', as a nuisance, as 'them not us', as the visitors who come and take up their time and want to be shown around, as family who think they (the migrants) are always on holiday (O'Reilly 2003).

On the other hand it has been undeniable in everything I write that this is a tourism-informed mobility (cf. Williams and Hall 2002). The migrants live in a tourist place, alongside tourists, sharing in tourist spaces and activities, spending time with tourists at work and play, often living leisured lives marked by tourist activities. British migration to Spain has been concentrated very much in the coastal zones which have been developed for tourism, with the infrastructure, tourist objects, activities, and services tourists are deemed to require. This argument has been pursued elsewhere (see O'Reilly 2003), but what continues to fascinate me about the Brits in Spain and sustained me in writing about this migration on and off for so many years, is the incredible contradictions that mark the migrants's lives, their identities, their experiences, dreams, aspirations, and frustrations. Their relationship to Spain is circumscribed by the fact that for them Spain symbolizes holiday and escape (and tourism), but they insist they are not tourists themselves.

---

1   I would like to dedicate this paper to Valerie, who very sadly and unexpectedly passed away since this paper was first drafted, demonstrating rather starkly the materiality of our bodies. She is missed, and I hope I have done justice to her sincere attempts to settle in Spain. Research for the paper was funded, between 2003 and 2005, by the Economic and Social Research Council of Great Britain (Grant No. R000223944).

2   There are currently 600,000 registered British living in Spain, and one can safely estimate that this only represents a third of those who actually live there most of the year. If all British in Spain were registered they would be the largest minority group in the country. See www.ine.es.

They declare a love of Spain while reminding each other 'we are guests here'. They say 'we live here' and yet go 'home' regularly. They say they want to integrate in Spain yet make little effort to put such aspirations into practice. They express a love of Spain but often seem to have a vague understanding of its culture and customs (O'Reilly 2000 and 2007, King et al. 2000, Rodriquez et al. 1998).

Yet even as I write these things I know I am making a lie of other things I have written, where I have argued that many British in Spain *do not* go home all the time, or that most *do* want to integrate and *do* try to settle (O'Reilly 2007). I have had to distinguish those who live in Spain all year round (what I have called full residents) from those who spend some of the year in Spain (seasonal, peripatetic or returning residents). And even then there are those who are simply experiencing life in Spain for a while and those who have committed themselves emotionally and financially and cannot easily return to the UK. There are some who spend a little time each year in Spain, but consider UK to be home, and others who live in Spain but regularly spend a little time 'back home' (O'Reilly 2000). A key problem with ethnographic research is that one becomes so familiar with individual people and individual lives it becomes impossible to continue to talk both generally and honestly about a group.

But however much I go round and round in circles trying to understand and explain these people and their problems and pleasures, intentions and impacts, it is impossible to deny their relationship to tourism and the way cultures of tourism impact on their lives. It is also impossible to deny the impact their cultures of tourism have on the surrounding areas, people and things. I am beginning to think that the term 'residential tourism', that many Spanish use to describe them, is actually a useful term after all, though I have avoided it because the migrants themselves argue against it. There is an extent to which all these migrants are turning tourism into a way of life and an extent to which the very contradictions outlined above are the very core of what they are and how they live.[3]

Things and Actions: Materiality and Performance

Usually I like to work inductively, entering the field, observing, joining in, and learning about people's lives from their own perspective before stepping out, seeing what sense I can make of it all for others, or how what I have witnessed and experienced can be translated or made sense of using theories and concepts to frame the various general themes I wish to explicate (O'Reilly 2005). However, for this chapter I have proceeded somewhat counter-intuitively. In the introduction the

---

3   There is an extent to which the argument in this paper is relevant for other North European migrants in Spain's coastal zones. However, research with these other groups has tended to focus exclusively on retirement migrants and to employ quantitative methods, so the applicability for this thesis to those groups remains a question (see Aledo 2005, Casado-Díaz et al., 2004, Rodríguez et al. 2005, Warnes et al. 2005).

editors of this book describe a new approach to tourism and culture that accounts for the relationships between people and things, the activities and objects of tourism, that understands the outcome of this interaction in terms of performance, and considers the impact of tourism cultures on objects and people. I decided to explore this new theoretical paradigm to see if there was anything that could act as sensitizing concepts that I might take back to the field and thereby 'see' things I had not seen before.

Haldrup and Larsen (2006) argue that materiality and material objects need to be incorporated into cultural accounts of tourism; that we need to re-examine the way things around us shape and are shaped by practice. This literature argues that the 'cultural turn' with its emphasis on thinking, imagining and representing, treats things as signs rather than concrete objects; whereas actions are only actionable because of the existence of things and technologies. They suggest that the world of leisure is not merely a human accomplishment, and seem to imply that *things* also accomplish. It seems to me this argument is a bit like Emile Durkheim's about the real nature of society, by saying people make and use things but then things come to have a life of their own – inasmuch as some things are doable and others not. I think this is partly true. Modern telecommunications technology, for example, has led to certain forms of communication, and therefore social life, that could not have been imagined in the past. On the other hand, mobile phone companies, for example, did not foresee the extent of the use of text-messaging and the impact that has had on young people's lives but have since adapted their technologies to take this into account, adding photo messaging and emoticons. This argument is not to deny that things have 'sign-value' but to acknowledge they also have 'use-value'.

In his fascinating introduction to the field of tourism, Adrian Franklin (2003) contends that tourism's objects can have a life of their own or can impact in unpredictable ways. John Urry's (1990) important attention to the tourist gaze, he believes, does not go far enough. Tourists do not just look upon, they buy, carry, transport, experience and use objects. In order to illustrate the role of objects, Franklin describes R. Ewins's study of bark cloth souvenirs in Fiji. Bark cloth, or tapa, was originally made from native plants and used in Fiji for clothing, bedding and ceremonial purposes. It was ornately decorated and also light to carry so was an ideal souvenir for the early tourists who arrived by steam ship. As a result locals started to make smaller pieces specifically for this market. However, although it has been common to argue that tourism thus commoditizes, trivializes, and ultimately extinguishes local cultures, Ewins argues that this market in tapa has actually ensured the continuation and even intensification of a tradition and economic activity which might otherwise have been lost as a result of economic and cultural competition from migrant Indian businesses.

These themes are not restricted to tourism studies by any means. In social anthropology, Tim Ingold (1995, 2000) has been developing an argument that things evolve as the outcome of the interrelationship of the human and non-human worlds; that the design of a thing is not so much driven by the human imagination as by its innate properties and malleability. Humans, by working on the world become part and parcel of the transformation of the system of relations within which both the human and the thing come into being. Here natural selection and cultural selection work hand in hand, so that like genetic dispositions, ideas are weeded out, adapted and developed over time and through the mutual weaving of people and materials.

In philosophy, Bruno Latour (2000) has been arguing that all human society is built with things. This is not simply to say that the material matters but to acknowledge that in fact there is no material versus human; humans are material and the material, as soon as it enters the social world (by being seen, thought about, used or remembered), ceases being an object. Rather than objects and subjects, for Latour there are only trajectories and transformations, paths and trails. Things take on the impression of our use, and social relations become mediated by things. But here things are *mediators* not intermediaries. In other words the things make or form social relations. However, at the same time these things remain frail, amendable.

Parallel to the development of ideas outlined above, others have been calling for more attention to the dynamic nature of place construction, to acknowledge change, fluidity and creation. Simon Coleman and Mike Crang (2002) argue that, contrary to the way they are portrayed in earlier tourist literature in which place = culture = people, and where tourists are often depicted as the cause of breakdown of this healthy relationship, places are in fact fluid, dynamic, multicultural, and created through performance. Again, these authors acknowledge the role the concept of the tourist gaze has played but recognize its limitations in looking only at what is seen, ignoring what is done and what is seen and done back. It is through embodied performance, they contend, that places (and lifestyles and cultures) are effected. They argue for the telling of spatial stories. 'We want to open up the possibilities of understanding tourism as an event that is about mobilizing and reconfiguring spaces and places, bringing them into new constellations and therefore transforming them.' (Coleman and Crang 2002: 10).

What I take from all this is the need to rethink what people do and with what on the Costa del Sol, and to ask what is the role of things, bodies and performance in residential tourism. For the rest of this paper I will respond to the call to consider the dynamic relationship between people and things, and the ways in which people actively and bodily involve themselves in the material world. I concede I have been guilty of seeing tourism purely as a system of signs, adding meaning to people's lives through what it signifies (pleasure, escape, contact with the Other, and so on). I have ignored the material in favour of the ideal factors in enabling and promoting first tourism then residential tourism, and neglected to consider residential tourism not as a state but as a continual performance, a balancing act of here and there, utopia and reality, tourism (travelling) and arriving.

The next two sections of this paper explore the material factors that have enabled and promoted tourism-related expectations and activities, thus incorporating materiality into the cultural account (Haldrup and Larsen 2006). Here I also acknowledge the way humans transform the world around them through their actions, and thus the co-creation of the material world and tourism culture in the Costa del Sol through transformation and performance (Coleman and Crang 2002). I then proceed to examine the complex material cultures that are woven into the everyday lives of British residents in Spain and demonstrate how it is the very contradiction between residence and tourism that enables residential tourists to live life as a kind of escape (sometimes less satisfactorily than others, see O'Reilly 2007). Residential tourism thus entails a subtle and continual balancing act, which is managed, practised or performed on a daily basis especially through the act of being both hosts and guests.

Materiality of Tourism

Regardless of Franklin's (2003) argument that tourism is not so much an escape from modernity as an aspect of it, I remain convinced that British residential tourism is driven by a desire to escape, a search for the authentic, for the past and for community. I am convinced of this because this is the way residential tourists frame their motivations. They continually disparage Britain (which is unusual for migrants) and illustrate the pull factors that attract them to Spain. According to a whole range of studies (Ackers and Dwyer 2004, Aledo 2005, Casado-Díaz et al. 2004, O'Reilly 2000 and 2007, Rodríguez et al. 2005, Warnes et al. 2005) their main reasons for moving are: for 'quality of life', the pace of life, or for a slower, relaxed life; the climate/ sun (which enables health and relaxation); the cost of living, cheap property (enabling early retirement and/or a better lifestyle); a business opportunity (to fund a better life); for a better life for the children; the culture (which they believe includes a sense of community, respect for the elderly, safety, and less crime); closeness to home, and other ties and connections; the desire to leave their home country (because of high crime rates, and too many immigrants(!), or to escape the rat-race, failing businesses, unemployment, or the political situation); and to go somewhere 'you can be yourself'.

One could argue that the Costa del Sol provides all these through imaginings, representations and the social construction of space (O'Reilly 2000). But it is undeniable that first of all the Costa del Sol was able to provide many things *materially*. The geographical area has innate properties humans work with (Ingold 2000). After all it has not been called the sunshine coast for nothing: it boasts an average of nine hours of sunshine a day. The sea is clear, blue and warm and the sands are golden. The heat of the summer especially, combined with the relatively low humidity mean no one goes anywhere in much of a hurry and the Spanish tendency to worry about things tomorrow rather than today is more in evidence in Andalucía than the rest of Spain. The coastal fishing villages and the surrounding white villages, symbolic of traditional unspoilt community life, were idyllic locations in which to locate the developing mass tourism of the late 1960s. Trail-blazing travellers, intellectuals and artists had already visited and recorded their experiences for others, including, for example, Ernest Hemingway and Gerald Brenan and other 'professionals' from diverse parts of the world (Fraser 1973). Tourists then merely followed these well-worn paths to established destinations (MacCannell 1976), seeing how others have used these spaces, going there and thus transforming them again in the process.

Of course, the gradual increase in visitor numbers together with the advent of the package holiday led to an unprecedented period of development in the 1970s. Some would say the area was spoiled, but for the tourists this also meant that materially they could still go to warm seas, golden sands, white villages, Mediterranean food and wine, and now with a built in infrastructure that made the whole thing easier, and it was still cheaper than home. Tourism theories that describe visitors sampling a little bit of the culture and the way of life without having to become too adept at the culture, without learning the language, and so on, really made sense in this situation (Boorstin 1964, MacCannell 1976). Indeed one could describe Costa del Sol tourism as the archetypal mass seaside tourism of the '60s and '70s. The Costa del Sol, I would argue, has been created for and by travel, tourism and now residential tourism more than any other tourist space. And of course, these types of mobility in turn have emerged out of the use and creation of this space for these purposes (Haldrup and Larsen 2006). Even Coleman and Crang have to agree in the midst of their argument against such a perspective, that 'if one observes the sprawl of concrete along the Mediterranean coast with its assorted "authentic English pubs", the vision of tourism as homogenizing and destroying local particularity might seem to have some credibility' (2002: 2).

Materiality of Residential Tourism

Materially, things have continued to develop and change in both the UK and in Spain. In 1975, the death of Franco removed many disincentives to migration for Europeans. Relations between Britain and Spain gradually became more Europeanized culminating in the free movement of individuals introduced in the Maastricht Treaty in 1992. The growth of the annual holiday combined with an increase in expendable wealth for many Europeans during the 1970s and 1980s, especially relative to Spain (Bedarida 1991), and owning a second home became the ultimate consumer aspiration in the newly affluent societies (Svensson 1989). A massive growth in the UK property market in the 1980s meant people could sell homes at huge profits and buy upmarket and cheaper in Spain, thus funding a new lifestyle in the bargain and, even during the subsequent property crash and downturn, the exchange rate between the pound sterling and the peseta remained favourable to Brits. Later, rural and inland tourism, and then what several Spanish academics and commentators have already called residential tourism (here referring to second home ownership and long-term tourism, Aledo and Mazón, 2004) were actively encouraged by Spain's tourist board to compensate for the seasonal and regional nature of tourism. This active promotion of property purchase and of tourism to rural areas has gone some way to encouraging the rural and inland migration of Europeans during the 1990s, thus taking on something of a life of its own (Franklin 2003). Costa del Sol residential tourism is both co-created and co-creating. Second-home owners were joined by retirees, looking to spend their third age (at least some of the year) in comfortable and reasonably priced surroundings. These were joined by bar and restaurant owners and others who provided services for the more settled migrants and tourists. As time has gone by younger migrants with families have joined the earlier migrants and now British are living in Spain in all manner of flexible ways. All this has been rehearsed elsewhere (King et al. 2000, O'Reilly 2000, Casado Diaz et al. 2004). I have reiterated it here in acknowledgement of the relevance of material (as opposed to merely ideal) factors in residential tourism.

Of course there are also more general material conditions which impact on residential tourism. We can summarize these as follows: globalization, increased interconnectedness and time-space compression (Giddens 1990); increased and turbulent forms of geographical mobility and a general fluidity in contemporary lifestyles (Papastergiadis 2000, Urry 2000, Bauman 2000); the rise of information technology and the network society; the spread of mass communications, cheap and expanded air and road travel; flexibility in labour markets, the end of the 'job for life', more working from home, the ability to live and work in different places, and with these, increased leisure time in affluent societies, extended holidays, and early retirement (Pollert 1991, Michie and Sheehan 2003). On a more negative note, increases in redundancies, temporary work, insecurity, and struggling small businesses encourage risk-taking behaviours such as working cash in hand (O'Reilly 2007); and finally migration chains, and the role of intermediaries – estate agents, financial institutions, mass media – promote and enable migrations (Castles and Miller 2003).

Utopian Dreams

Increasing numbers of residential tourists, then, are attempting to emulate the tourism-informed lifestyle in more and more flexible ways. The one thing that has remained continuous is the description of the place they are going to as escape, even as materially things have changed. But not only do they describe escape, they live and perform it on a daily basis. They tend to avoid routine, to focus on leisure rather than work, to work informally or casually, and to deny routine pressures and strains. They spend a lot of time with holiday-makers and thus avoid serious conversations, and they are loathe to have their lives structured or controlled, often avoiding registering with the town hall or paying taxes (O'Reilly 2000 and 2003). In other words their lifestyles and (material) cultures are *effected* in relation to material factors through embodied performance (Coleman and Crang 2002).

MacCannell (1976) famously defined tourism in terms of the existential quest for the authentic, arguing that as modern life becomes increasingly adulterated the desire for the real or original increases. Now, many tourism theories have challenged this perspective, but only in as much as it is based on an assumption that there is a real, authentic or original to experience. Authenticity can only be staged if there is such a thing to stage, and preserve (Coleman and Crang 2002). Franklin (2003: 24) goes further, arguing that tourism is not so much a desire to escape modernity as a means to take part in it; 'the quintessential expression and performance of modern life'. But I think even if we accept (as we must) that cultures are not bounded, coherent or firmly located, and that tourism is as much part of the everyday as an attempt to escape it, nevertheless we must still acknowledge that British residential tourists describe a search for escape, community, the past, the real, the slow, and the uncomplicated.

I do not think that this escape is necessarily an escape from modernity or from urban lifestyles (that assumes that life actually was once better); I think it is simply that there is a timeless idea that life was once better, or that it could be better. The themes are the same as far back and as broadly as the timeless theme of utopia. John Carey (1999) has traced utopia as a genre as far back as ancient Egypt. Utopias can be forecasts, dreams, stories, descriptions, the imaginary voyage, the earthly or heavenly paradise, the political manifesto, and of course the search for Eldorado. Carey believes utopias express a desire to replace the world that we know, whatever that world looks like, but they do share key themes which we can certainly recognize in residential tourism: the desire to get rid of criminals (or to move away from them), a search for selflessness and community spirit, a search for quality of life, tranquillity and beauty (and sometimes achieving this by looking inward and ignoring the world around you). Utopians are fair and reasonable and true to 'man's' true nature (whatever that is taken to mean), and utopias often look back to the past or forward to a better future, and include transit to some other place, 'where something can be learned about how life should be lived' (Carey 1999: 1).

But utopia is nowhere. It cannot exist in reality. Tourism works because (for some types of tourism) it is about seeing this other world for a while but then going back to reality; gazing on it but staying safe. Residential tourism works as long as the participants remain in but not in, of but not of, home but not home, neither here nor there. In other words the contradictions are essential: they explain the lifestyle. Residential tourism is managed by continually balancing tourism and residence, and one of the key ways this is achieved is through acting as hosts to their guests. It is fascinating to see this in action. Residential tourists learn all about the area, show their visitors around, share their knowledge, and act in many ways as tour guides. At the same time, of course, they act as tourists themselves, joining in activities and gazing on sights and landscapes. The act and performance of touring, showing people around, taking part in tourist activities marks them as both hosts and guests.

Hosts and Guests

In 2001 Mike and Valerie, who were both then in their 50s, decided to quit their jobs, sell their house and move to Malaga for a better life. They calculated that if they sold their house in the UK and bought a cheaper one in Spain they could live carefully off the remainder of their capital until they reached retirement age and could claim small pensions. They bought a house in the countryside, partly because Mike wanted a big garden to work on and partly because they could not afford anything very close to the coast. They spoke very little Spanish before they arrived but had every intention to learn the language once they had settled. To an extent they assumed that living in the countryside would mean they would learn the language simply by being there and having to use it. In fact being early retired meant that their contact with Spanish people was restricted to shop assistants and waiters and the occasional health or council official. They found that it was not at all easy to just go out and get to know Spanish locals. On the other hand, given that there are many British people living in the area, it was very easy to meet their own compatriots, make friends, get help and advice, and generally build up a good network of friends and acquaintances who all spoke English. This is not what they had intended, of course, it just kind of happened. They continue, even now, to attend Spanish classes regularly, and Valerie has an arrangement with her dentist's assistant whereby they meet regularly for Spanish and English conversation. But Valerie is sad that the friendship has not developed further and that they seem to have made no real Spanish friends, other than people to simply nod and say 'hello' to in the neighbourhood.

I have explained this lack of integration as being the result of a combination of factors: many British in Spain are retired and therefore do not meet Spanish people in their daily lives in order to build relationships and acquire good language skills; many of those who do work, work in the tourist industry or with other migrants (in estate agencies, bars, timeshare sales, and property maintenance); there are so many tourists and other migrants in the area it is easier not to speak Spanish than to learn it; and as more younger migrants settle in the area the more they provide work for themselves in the form of businesses offering a range of services to both the settled migrant community and the tourists. It is now possible to get your hair cut and styled, your dog's hair cut and styled, have an extension built on your house, have your garden maintained, buy a car, get your pool cleaned, and have your legs waxed without speaking a word of Spanish (and perhaps without even meeting a Spanish person). On top of that one can receive British or English-language daily papers and even watch, through satellite or cable, all the British freeview or English-language television channels.

Mike and Valerie spend their time engaged in a multitude of activities. They are members of a gardening club and an amateur theatre group, Valerie does line-dancing and runs a small brownie-guide club, they play tennis and are members of a bird-watching group. All of these groups are English-speaking and all are made up of a mixture of (mostly English) migrants, seasonal migrants and sometimes tourists or visitors. However, one activity they engage in above all others is being hosts to their long stream of visitors from the UK. Like many of the British who are settled in Spain they complain about the cost, effort, strain and interruption to their routine caused by the fact that one can ensure above all else that when you live in Spain people will visit you. In a survey I conducted amongst those who live in Spain 97 percent said they have visitors from home and some stay up to a few months at a time. As one man put it half jokingly:

> It comes to something when you've got to move two thousand miles to be popular, but all of a sudden all these friends that I haven't spoken to for years are suddenly, "oh can we come and stay with you for a week?" And I go "sorry, who are you again?" and they say something like "Oh, I was behind you in the queue in the bank once. You know me!"

The contradiction is in the fact that although they complain about it, they do it all so well and even seem to encourage the activity. I would consider visiting friends and relatives to be a material fact of living in Spain for British migrants, and the way this is managed can be understood as a performance of the balance between hosts and guests, between living and touring, being of but not of the society they live in.

Valerie and Mike assiduously collect tourism leaflets and brochures and general information about their area to the extent that they would make tour guides appear uninformed. Valerie knows where the tourist office is in her town (though it took me weeks to locate it), and regularly picks up the newspaper, *Malaga Rural*, which richly describes local villages and events. She is familiar with the location of several national parks and public gardens in her area. She knows exactly when what ferias are on in which village, likes to go with visitors to as many as possible, and is able to explain the meanings of a range of religious ceremonies and processions. She knows the date of *Coín's Festival de los Naranjos*, and the best spot to witness the celebration of the *Noche de San Juan*, which takes place in coastal towns on the longest day of the year. She loves to take visitors to *Mijas's International Day*. Mike knows the fastest or most scenic routes to historic and other tourist sites, the viewing places to stop at *en route*, and the location of the best car parks. He can take visitors swimming in the reservoirs at El Chorro Gorge and to see the place where *Von Ryan's Express* was filmed, in the space of one day, passing a café on the way there, a restaurant at midday, and a café on the way back. Valerie and Mike have located Roman baths in Antequera, know when to spot griffon vultures in El Torcal, and enjoy taking visitors on the costal walk from La Cala del Moral.

Valerie and Mike believe that learning about local places and events, histories, geographies and cultures, makes living in Spain more interesting for themselves. They explain that having many visitors throughout the year can become tedious without new places to visit. Also, visitors tend to return more than once and so need new places to see themselves. In this way Valerie and Mike are merely being good hosts, sharing and describing their home. However, being retired and living leisured lifestyles, they are thus free to join their tourists in their touring and so, like many of the migrants in Spain, they try to treat the time when their visitors come as a bit of a holiday and sight-seeing tour themselves. To this extent they are tourists.

But Valerie, Mike and their visitors are never really part of what they see and do. For a start locals do not go to lots of ferias and fiestas, just their own, especially when it comes to the *Romería* or the Easter parades. And when residential tourists and their visitors stand, as I have done many a time, watching a procession or a *Romería*, or taking part in a giant country picnic, they tend to feel out of it, lonely while in the thick of things, left out. They remain strangers, for whom nothing can be taken for granted and all is open to question. As Mike explained: 'you can read all you like about the different things, but you are never really sure what is really going on, and you always feel like an outsider really'. For all migrants, the world around them is a field of adventure and a topic of investigation (Schutz 1971) and Mike's and Valerie's studying and attempts to take part are merely more of that. The more they learn and try to engage the more like strangers they become.

Valerie and Mike, like many migrants in Spain, also complain about being hosts to so many guests. It can get boring and tedious, and costly, especially if people stay too long. And, after all, the tension between trying to live somewhere and trying to be on holiday is experienced as any tension, and is never quite comfortable. They tend to eat out a lot when visitors come, and women thus complain that it is impossible to maintain a healthy diet. They spend a lot of money on fuel taking visitors around. Mike says he sometimes feels like a taxi-driver. Valerie gets worn out with cleaning the house in preparation for guests who treat her home like a hotel. They want to continue to do their routine things – like gardening and learning Spanish – and have to fit these in around tourist activities. But this performance of hosts and guests is the daily practice of balancing tourism and residence and cannot be resolved.

Another aspect of this balancing act between home and away, touring and residence is that many British migrants in Spain return to the UK regularly, where they are reminded they are now away from home in what used to be their 'normal' surroundings. My survey of 340 British residents was only distributed to those who consider they live in Spain and yet, nevertheless, 64 percent of respondents return to the UK each year and a quarter of those had returned for more than one month in the past year. Mike and Valerie go back each summer, to escape the intense heat in Andalucía at that time of year. Valerie goes for a couple of months. It gives her time to see her mother and her grandchildren and to escape what she calls the trials and tribulations of living in Spain: the surprising cold and damp in the winter, the frequent power cuts, the way her access road gets flooded each winter, the cracks that are appearing in her bedroom walls as a result of extreme weather conditions, the stray dogs that drive her mad with their barking at night, and constant frustration at her lack of Spanish language skills. Mike goes for just a few weeks as he enjoys working on his garden in Spain and hates to leave it too long. Also, he feels too much like a visitor and thus a nuisance in the UK. Valerie, on the other hand, feels a little bit as if she is going home again each time.

Conclusion

In conclusion, it seems residential tourism (a term I have avoided over the years) might be the one best suited to explain British migration to Spain's coastal areas. Residential tourism is an oxymoron, a contradiction in terms, and British residential tourists practice, perform and experience this paradox on a daily basis. Recent developments in philosophy, anthropology, cultural studies and tourism that focus on the inter-relationship of people and things, that explore actions and the performance and co-creation of cultures and places, help us understand residential tourism as the daily practice of a contradiction. Residential tourism is an escape attempt, a utopian dream, and Spain's coasts materially offer idyllic locations to which to escape. The creation of these spaces as sites for pleasure and leisure has then provided the infrastructure that enables residential tourism. Later, supported by a host of material factors, the settled tourists themselves further facilitate the migration of others. But utopias are nowhere and cannot exist in reality, so the escape attempt has to be managed on a daily basis by balancing competing fields (home/away, tourism/residence). One of the most stark ways this is witnessed is through the practice of hosts and guests: residential tourists are both.

References

Ackers, L. and Dwyer, P. 2004. Fixed laws, fluid lives: the citizenship status of post-retirement migrants in the European Union. *Ageing and Society*, 24(3), 451–75.

Antonio-Tur, A. 2005. Los otros inmigrantes: residentes europeos en el sudeste español, in *Movimientos Migratorios Contemporáneos*, edited by J. Fernández-Rufete and M. Jiménez. Murcia: Fundación Universitaria San Antonio, 161–80.

Bauman, Z. 2000. *Liquid Modernity*. Cambridge: Polity Press.

Bedarida, F. 1991. *A Social History of England 1851–1990*. 2nd Edition. London: Routledge.

Boorstin, D. 1964. *The Image: A Guide to Pseudo-Events in America*. New York: Harper.

Carey, J. 1999. *The Faber Book of Utopias*. London: Faber and Faber.

Casado-Díaz, M.A., Kaiser, C. and Warnes, A.M. 2004. Northern European retired residents in nine southern European areas: characteristics, motivations and adjustment. *Ageing and Society*, 24(3), 353–381.

Castles, S. and Miller, M. 2003. *The Age of Migration*. 3rd Edition. Hampshire and New York: Palgrave Macmillan.

Coleman, S, and Crang M. 2002. *Tourism: Between Place and Performance.* Oxford: Berghahn.

Franklin, A. 2003. *Tourism: An Introduction.* London: Sage.

Fraser, R. 1973. *The Pueblo. A Mountain Village on the Costa del Sol.* London: Allen Lane.

Giddens, A. 1990. *The Consequences of Modernity.* Stanford, CA: Stanford University Press.

Haldrup, M. and Larsen, J. 2006. Material cultures of tourism. *Leisure Studies*, 25(3), 275–89.

Ingold, T. 1995. Building, dwelling, living: how animals and people make themselves at home in the world, in *Shifting Contexts*, edited by M. Strathern. London: Routledge, 57–80.

Ingold, T. 2000. Making culture and weaving the world, in *Matter, Materiality and Modern Culture*, edited by in P.M. Graves-Brown. London: Routledge, 50–71.

King, R., Warnes, T. and Williams, A. 2000. *Sunset Lives: British Retirement to Southern Europe.* Oxford: Berg.

Latour, B. 2000. The Berlin key or how to do words with things, in *Matter, Materiality and Modern Culture*, edited by P.M. Graves-Brown. London: Routledge,10–21.

MacCannell, D. 1976. *The Tourist: A New Theory of the Leisure Class.* New York: Schocken.

Michie, J. and Sheehan, M. 2003. Labour market deregulation, 'flexibility' and innovation. *Cambridge Journal of Economics*, 27(6), 123–43.

O'Reilly, K. 2007. Intra-European migration and the mobility-enclosure dialectic. *Sociology*, 41(2), 277–93.

O'Reilly, K. 2005. *Ethnographic Methods.* London: Routledge.

O'Reilly, K. 2003. When is a tourist? the articulation of tourism and migration in Spain's Costa del Sol. *Tourist Studies*, 3(3), 301–17.

O'Reilly, K. 2000. *The British on the Costa del Sol.* London: Routledge.

Papastergiadis, N. 2000. *The Turbulence of Migration: Globalization, Deterritorialization and Hybridity.* Cambridge: Polity Press.

Pollert. A. 1991. *Farewell to Flexibility.* Oxford: Basil Blackwell.

Rodríguez, V., Fernández-Mayoralas, G., Casado-Díaz, M.A. and Huber A. 2005. Una perspectiva actual de la migración internacional de jubilados en España, in *La Migración de Europeos Retirados en España*, edited by V. Rodríguez, M.A. Casado-Díaz and A. Huber. Madrid: CSIC, 15–46.

Rodríguez, V., Fernández-Mayoralas, G. and Rojo, F. 1998. European retirees on the Costa del Sol: A cross-national comparison. *International Journal of Population Geography*, 4(2), 91–111.

Schutz, A. 1971. The stranger: an essay in social psychology, in *Alfred Schutz: Collected Papers II: Studies in Social Theory*, edited by A. Broderson. The Hague: Martinus Nijhoff.

Svensson, P. 1989. *Your Home in Spain.* 2nd Edition. London: Longman.

Urry, J. 2000. *Sociology Beyond Societies: Mobilities for the Twenty-first century*. London: Routledge.

Urry, J. 1990. *The Tourist Gaze*. London: Sage.

Warnes, T., King, R. and Williams, A. 2005. Migraciones a España tras la Jubilación, in *La Migración de Europeos Retirados en España*, edited by V. Rodríguez, M.A. Casado-Díaz and A. Huber. Madrid: CSIC, 47–68.

Williams, A. and Hall, C.M. 2002. *Tourism and Migration: New Relationships Between Production and Consumption*, London: Kluwer Academic Publishers.

Chapter 8

Mobile Practice and Youth Tourism

Dan Knox

Introduction

This chapter is concerned with the relations between everyday life and mass tourist practice, and the ways in which mass youth tourism could be said to echo accounts of cultural tourism as consumption of familiarity (Prentice and Andersen 2007). The illustrative case-studies of clubbing on holiday, holiday sex and the taste of home are about the signifying practices of those participating in clubbing holidays to Ayia Napa in Cyprus and Faliraki on the Greek island of Rhodes. It is also about both prior and subsequent written, visual, filmic and musical representations of such holidays, and how the circulation of those representations fuels the continued consumption of youth tourism products as well as the condemnation of cultural commentators, the media and the general public. To date, the hegemonic paradigm of tourism research has been concerned primarily with the spectacular and the 'Other' as the object and motivation of tourism behaviours (Urry 1991). This project turns instead to look at the everyday, the banal and the familiar, and builds on earlier work by the same author (Knox 2006, 2008) in responding to the problematization of the distinction between the 'holiday' and the 'everyday' (Franklin and Crang 2001, Stebbins 2001). Such a conceptual move to the consideration of the quotidian is inherently bound up with broader interests in performativity and practice seen across the social sciences and humanities (Hannam et al. 2006, Knox 2006, 2008), as well as in tourism research (Bærenholdt et al. 2004, Coleman and Crang 2001, Crouch 1999, Mordue 2005). This project is concerned with the relationships and intersections between the 'spectacle' (Baudrillard 1990, Debord 1967) and the 'banal' (Billig 1995, McKee 1997, Palmer 1998) and the processes through which both are created and sustained.

The research on which this chapter is based was concerned to detail the practices of clubbing tourists and the relationships of these to the rhetoric of tourism and leisure providers in Ayia Napa and Faliraki. The research has established, through qualitative content and discourse analyses, the most significant features of relationships between garage music and the iconographies of youth tourism to show that representations of the two resorts in the United Kingdom exercise a strong regulatory power over the practices of tourists.

The central premise of the research has been to query where everyday practice ends and holidays begin, and as such attention has been focused on the mundane aspects of tourism. These apparently banal aspects of clubbing holidays include practices very much like those we can observe in other mass tourism settings – the consumption of alcohol, lounging around the poolside, partying and dancing, and the eating of characteristically home-nation cuisines. The research has been conducted through extensive participant observation during the summers of 2003 (Ayia Napa) and 2004 (Faliraki), content and discourse analyses and interviews with young British holidaymakers in both destinations. The primary data source was collected in a detailed research diary, recording observations, the experiences of participation and opportunities that being there and taking part provided for speaking to tourists. Participant observation was conducted primarily in the hours between the late afternoon and 7 am in settings such as nightclubs, bars, restaurants and takeaways, hotels, beaches and on the streets. The range of materials examined for content analysis was collected in the UK, in Greece and in Cyprus and includes television shows, newspaper and magazine articles, videos, music, promotional literature, holiday company brochures, club flyers and websites.

UK Garage, Ayia Napa and Faliraki

Since the late 1990s, Ayia Napa, a small town in the south of Cyprus, has become one of the key destinations for young Britons seeking clubbing and dance music centred holidays, as well as catering to the more mainstream mass tourist. Youth and clubbing tourism to Ayia Napa is particularly focused on the UK Garage sub-genre of electronic dance music, in contrast to both Ibiza and Goa where House and Trance music respectively thrive (Salahnda 2002). The Greek resort of Faliraki, near to Rhodes town, has similarly been repositioned as a clubbing and youth holiday destination over the same period of time, and there has been something of shift away from Ayia Napa towards Faliraki for some elements of the UK Garage music scene. In some regards, this geographical shift can be seen as a move away from a place that had become troublesome in terms of violence and policing, as well as an attempt to maintain the authenticity of the experience for the core of scene members. As ever greater numbers of non-enthusiasts began to attend the garage music events it was felt that something of a sense of community had been lost.

Some researchers (Malbon 1999, Skelton et al. 1988) have begun exploring the social performances of the UK club scene, and many others the politics and embodied experiences of dancing (see Nash 2000) but thus far very little work has been undertaken that examines clubbing or youth holidays abroad. Tourism researchers have explored the destinations and motivations of gay holidaymakers (Clift and Forrest 1999) and the relations between sex and tourism more generally (Clift and Carter 2000, Opperman 1999, Ryan and Hall 2001), while Sellers (1998) and Saldanha (2002) have explored the connections between tourism and clubbing, and in particular the practices of clubbers. The exposition of the intertwining of tourism, clubbing, sex and hedonism is taken further in this research by embedding these practices within a broader and deeper social and cultural context that takes account not only of everyday life at home, but the totality of tourist practices enacted.

Tourism, Sex and Hedonism

Surprisingly, it remains the case that there is a notable lack of research on youth holidays both in tourist studies and in the broader social sciences. This is despite the above noted wealth of material concerned with youth cultures, popular cultures, and subcultures including clubbing and dancing. A principle area in which hedonism has been explored has been in writing about adventure tourism and/or extreme sports, with fun, sex, drugs, drink, dancing and violence attracting less attention. There is an expansive literature on tourism and sex that has spent an inordinate amount of time defining itself in relation to what we might, or indeed might not, call 'sex tourism'. This literature explores the distinction between consensual/non-consensual and commercial/non-commercial sexual behaviours but has largely maintained a focus upon the more negative aspects of the intersections between tourism and sex. Thus, the non-consensual and commercial practices of a variety of modes of prostitution have been to the fore of such debates, partly as a result of the political motivations of researchers, and partly as the result of failure to expand analyses. Similarly, much attention has been focused on the specifics of gay tourism as the public face of homosexuality on holiday. Gay tourism revolves around the conscious self-identification of travellers as gay and is largely about travel to what have come to be seen as gay destinations or gay-friendly cities such as Amsterdam or San Francisco. This is one of the more intriguing elements of gay tourism; the ways in which it appears to be built upon established networks of people and so-called gay spaces or places. It is suggested that gay people are peculiarly disadvantaged in that they only truly have the freedom to self-identify as gay in particular places at particular times. This issue ought to be of particular interest to all academics with an interest in tourism, as it runs somewhat counter to more traditional arguments about the holiday being an opportunity to transcend everyday life and identities. It could seem that gay holidaymakers are in some sense better served than heterosexual companies in that there are noted destinations in which aspects of homosexual sexualities are invited, welcomed and celebrated, whereas heterosexual supposed-norms attract censure when they manifest in the ways that they do around the Mediterranean.

Understanding Mass Tourism

The mass tourist is often vilified as uncultured, uncaring and undeserving. Despoiling natural environments, producing photo-fit resorts and lacking the interest in local cultures and traditions that would enable them to distinguish themselves within social groups, mass tourists are seen as unthinking drones, flocking to whichever resorts are most effectively marketed in their direction by large travel corporations. While there may be some elements of truth in these kinds of accounts, what I want to do here is to make a passionate plea for a deeper and more nuanced understanding of the behaviour of mass tourists in order to redress the balance. Sneering at people we do not understand is all very well, but sadly fails to offer a significant insight in the same way that conducting ethnographic research among them might. In fairness to the middle-class scholars and researchers of tourist studies, it is not so much that they have tended to view mass tourists negatively so much as they have failed to notice very much about them at all. Certainly, we could argue that the Spanish coastline has become relatively unattractive in relation to an idealized notion of pristine nature, but it would be a mistake to try to argue for effective environmental solutions without an understanding of what motivates people to continue to visit such apparently unattractive places. It may well be that such places have an attraction that tourist studies have thus far failed to grasp in any meaningful sense.

Bourdieu's (1983) account of the gathering of cultural capital by the new middle classes has provided some interesting conceptual material for those working within tourist studies to explore the creation and maintenance of distinction among elite cultural tourists. While notions of cultural capital apply most obviously to middle class social groupings, I want to argue that very similar processes relate to the patterns of conspicuous consumption of mass tourists. Building a tourist biography that takes in Prague, Rome, New York, Barcelona and Krakow, incorporating the Sistine Chapel, the Louvre, the Guggenheim and the Ramblas, enables the tourist to accrue layers of cultural capital that function either as shared experiences or aspirational goals for other members of their social grouping. Similarly, a lifetime spent visiting Faliraki, Magaluf, Ibiza, Ayia Napa and Kavos, and experiencing the tourist products of those places, functions as a marker of social distinction within the social groups of the mass tourist. Some among this group might accidentally find themselves on a weekend break to Prague, but will usually ensure that no conventional cultural capital is earned during such a visit by avoiding the galleries, museums and castles in favour of the Irish pubs, the English pubs and the strip clubs. To criticize such visitors on account of their non-participation in middle-class cultural tourism is to ignore the value of a more hedonistic visit to Prague because of an entrenched view of what kinds of tourism should be conducted where.

Liminality

The total range of practices that constitutes a two week vacation in Faliraki or Ayia Napa is entirely contained within a very neatly delineated and limited space. This space is the destination resort and is easily identifiable as matching the limits of large hotel developments, after which the countryside fades into scrubland and seemingly more authentic local residential properties. Not only is there no need, there is no incentive for the average youth tourist to explore beyond these boundaries as all the drinking, partying, dancing, fighting and sexual opportunities are concentrated within the tourist resort. It is not that in different circumstances such tourists might not be interested in visiting other kinds of attractions, but merely that the kinds of holidays on which they have embarked are designed to facilitate lounging by the pool, drinking and attempting to sleep with similar members of the opposite sex.

The concept of liminality is somewhat problematic in these instances as mass youth tourists are not necessarily suspended between home and away in a vacuum of what might be termed bad behaviour by British, Greek and Cypriot societies. It is equally valid to argue that such visitors have very definitely arrived in Faliraki while leaving behind the censure of home and avoiding the censure of Greece. The rules and conventions of behaviour in Faliraki are not those of the territorially more expansive Greek nation, nor indeed those of the United Kingdom, but something different and contingent on the intersection of location, guests and hosts. Transient and mobile communities establish rules of behaviour in much the same way that permanent, fixed communities do, but it may be more difficult to trace out the ways that people adhere to these where populations are subjected to around a 90 percent change in personnel every two weeks.

The arena for hedonism is sustained by those parts of the population that do not leave the islands until the end of the summer. Their ongoing behaviour serves to encourage newly arriving visitors in the debauchery and thrill-seeking that had already served as their motivation to visit. The owners and managers of venues, club reps, seasonal staff and local people all come together to ensure the smooth provision of the architectures of hedonism in an environment in which such apparently rule-shattering behaviour has become the norm. Once a context within which hedonistic behaviour has become established as ordinary and unremarkable practice has been instituted, such behaviour begins to look rather contrived, controlled and conventional. True rebelliousness within the context of Ayia Napa or Faliraki would require modest consumption and self-censure. In part this would constitute a refusal of the youth tourist resorts themselves that would negate the act of choosing such a destination.

Scene 1: Holiday Sex

During my research, it became apparent that self-censure and friendship group censure both operated in such a way as to encourage what could be seen as reckless or wild behaviour. The norms and conventions of behaviour at home very clearly no longer apply for many people, but they need not often apply for many of these groups of youths at home. Most people visited the resorts in same sex groups ranging in size from four to ten, and sharing two or three to a room in large hotel complexes. In such circumstances, the practices and processes of group surveillance are ever-present for each individual in even the most intimate and personal of circumstances. While it would not be unusual for friends sharing a room to both engage in sexual acts with new-found partners at the same time, consensus appears to be that such activity ought to be staggered in some way. Marking off shared spaces in this way has the effect of temporarily excluding some guests from their rooms while very short-term relationships are enacted. The act of separating out space in this way was justified by male tourists in terms not of providing intimacy or privacy but in terms of sparing roommates the embarrassment of being present at the same time and the potentially harmful effects of seeing another man's erect penis. Such rampant heterosexuality should not, as far as these rampant heterosexuals were concerned, not contain anything that might be misconstrued as latent homosexuality. Rachel, aged 21, from Coventry told me the following:

> I don't really want to have to watch the girl that I'm sharing a room with going at it with the bloke she picked up, especially if I haven't managed to pull someone myself, you know? I'll usually wait outside if she gets lucky, maybe have another drink, hopefully hang around with someone else until she's finished, however long it takes, haha (Interview 2004).

As with the males, the non-participation in somebody else's intimacy appears to be driven by an unwillingness to be exposed to such behaviour rather than the need to ensure privacy for the participants. As Callum, aged 19 from Glasgow, put it, 'there are some things I just don't need to see, no matter how good a mate someone is'. The hotel room acts as the provision of a space in which to have sex with recent acquaintances, where such a space is not available at home, has been recognized as an important part of the relationships between sex and tourism. It is not necessarily that people are engaging in more sexual acts during each night out than they would at home, but rather that the acts themselves are different in terms of duration and location. The sharing of a hotel room merely acts as a minor obstacle, but ordinarily an obstacle that is somewhat easier to overcome than the sharing of a family home with parents.

Scene 2: Clubbing on Holiday

Garage music in the UK underwent various transformations from the late 1990s to emerge in the early twentieth century as a distinctively British form of music concerned with (sub)urban life in the United Kingdom. The musical forms of Garage are diverse though some of the key elements that can be pointed to include the rhythms of drum and bass and two-step, and the vocal styles of Hip Hop, Ragga and contemporary American R n' B. Tied in with this content and these styles, Garage music has an association with violence and danger, and a reputation for the perpetuation of the gangster image as imported most recently from US Hip Hop. The crucial characteristics of a gangster image are violence, misogyny and criminality. UK Garage is a global form, but a global form with highly localized roots in the United Kingdom.

In addition to British roots, the UK Garage scene has a particular and established summer holiday destination in Ayia Napa. There is a blurring here between the everyday and the holiday in that large sections of a UK community remove themselves temporarily from home and reconvene as a community in a location in Cyprus to continue, indeed to intensify, the activities they partake of together at home. Indeed, it is as if the community itself only truly comes together each summer in Ayia Napa, in a way that is not possible within the United Kingdom, to stage club nights, attend parties and perform. Ayia Napa has become a focal point for an imagined community and a symbolically important space for the UK Garage community. A two-week holiday in Ayia Napa serves a purpose similar to that of a national festival or ceremony in maintaining a community of interest, and thus a distinction between the everyday and the not-so-everyday is retained. Quotidian domestic practice and experience can be seen to be preparing individual holiday makers for the experiences of a clubbing holiday to Ayia Napa that is not necessarily terribly different, other than in terms of location, to a weekend night out at home. Despite, in some sense, being rooted in Ayia Napa, these discourses and practices are both preceded and succeeded by similar practices in other spaces and at other times. Popular culture here is both mobile and international, with Ayia Napa being merely one node in a cultural network.

As Angus, a 23-year-old from Newcastle, told me:

> I have bought some of the Ayia Napa compilations, they're pretty commercial, but they sum up some of the scene here and at home. We've brought some of them with us to listen to on the beach and in the hotel – they're cool. We listened to them at home before we came here ... and you do hear the same tunes, MCs, beats, whatever....

The range and duration of drinking and clubbing opportunities offered to British tourists in Rhodes or Cyprus, combined with the context of being on holiday with a group of friends, provides numerous opportunities for the excessive consumption of alcohol, or drugs, to dance, to consume UK dance music, and to dance with other holidaying Britons. Despite the vaguely exotic setting, much about the clubbing scenes of Faliraki and Ayia Napa will be familiar to those acquainted with the British club scene, and more so for those with previous experiences of either the same resorts or of Ibiza. Indeed, the similarities to home are far more striking than the differences raising questions about the extent to which two weeks in either resort might realistically count as an escape from anything at all – it could in fact be argued that one of the worst things that you could do if seeking to escape home would be to take a holiday in a place that is likely to be frequented by many people from home, will feature club promotions and parties staged by UK clubs or DJs, include performances by UK-based DJs and MCs and enable repeat hearings of tunes, beats and samples familiar from home. Michelle, 26, from London stated that: 'So many things here are just like at home, so it's kinda just a big UK Garage party in Greece, it doesn't matter that we are in Greece, we could be in London, we could be anywhere'.

The degree to which similarities exist is partially contingent on the individual and visitors can be broadly divided into two groups. The first group, for whom the continuity with home is greatest, includes performers and active members of a UK Garage scene, while the second category includes all other people with perhaps only a passing interest in that specific sub-genre of dance music or more general hedonists simply looking for a good time. For the self-identified members of a UK Garage community, it is likely that not only are some of the promoters and performers already well-known to them, but also that they are familiar with others they find themselves dancing and drinking with in the clubs and bars. The more casual consumers, however, are not operating in an unknown world in that most of the practices of clubbing are very much the same as they would have experienced at home. For example, foam parties may not be weekly events in the United Kingdom, but they are far from unknown – all being in a Mediterranean holiday resort contributes is the possibility to attend such an event everyday or every second day. The co-presence of unknown fellow clubbers is, again, not unlike attending nightclubs in the UK in that even where people are strangers, a generally friendly atmosphere is maintained, with temporary friendships readily formed as a part of a shared experience. All that is required for the process of bonding with new friends is the shared interest in the music, the hedonism or the subculture, and this is readily evidenced by simply being there.

Scene 3: The Taste of Home

The shared experience and the shared backgrounds of people on clubbing holidays to Mediterranean resorts is strengthened by their use of a shared tourist infrastructure during the hours when they are not participating in dancing or drinking. A visit to the convenience stores catering to tourists in either Faliraki or Ayia Napa provides the British tourist with any number of comforting and reassuring brands and products. Naturally, truly global brands such as Coca-Cola or Budweiser are widely available, but in these spaces of mass tourist consumption are offered alongside imported Cadbury's chocolate, Golden Wonder crisps, Walker's shortbread, Kingsmill bread, Daddies sauce, Tetley teabags and so on. The provision of such authentically British foodstuffs admittedly does some work in terms of de-exoticizing Greece or Cyprus, reassuring tourists that they have not strayed too far away from home. To criticize tourists, however, for consumption of such homely products would be to misunderstand why they have visited Faliraki or Greece by viewing such encounters from the perspective of somebody with a different set of motivations. Youth tourist holidays are not about experiencing apparently unspoiled authentic cultures or attempting to go native in Greece or Cyprus, but rather about the opportunity to have more fun in a short space of time than might be possible at home. If British seaside resorts were as reliably warm as Cyprus or Rhodes, a proportion of the resorts on those islands would be conducted in Brighton, Whitley Bay or Margate instead. Thus, the act of sitting in a bar in Ayia Napa and ordering a full English breakfast, or some approximation of such, should not be seen as a failure to grasp the essential nature of Greek-Cypriot culture nor necessarily an act of conservatism intended to protect the tourist from the dangers of local cuisine. Neither, however, should it be read as an act of resistance against the mores of middle class cultural tourism or as a manifestation of common-sense in the face of increasingly difficult-to-sustain notions of tourist authenticity. Rather, such a consumptive act, undertaken in an English or Scottish or Welsh-themed bar, is partially about an assertion of national identity, but largely about the association of greasy bacon, eggs and sausages with heavy alcohol consumption in popular understanding: a large fried breakfast is widely-considered to have restorative effects on the hangover body as well as being seen to prepare the stomach for further alcohol consumption. Indeed, in many cases, the eating of such a all-day breakfast is often accompanied with the drinking of the first beer of the next day.

The taking up of the opportunity to eat a roast dinner, drink a pint of British beer and watch *EastEnders* on satellite television is as much a manifestation of disappointment with Britain itself as it is with Greece or Cyprus. Rampaging, or otherwise, 'Brits on the Piss' would not be found littered around the shores of the Mediterranean in such numbers if the UK offered the same likelihood of exposure to the shining sun. However, where there is a specific interest in a particular social subcultural scene, the motivation to locate temporarily in these resorts is more complex, although we would be hard pushed to understand this from media representations that fail to differentiate groups of youth tourists.

Representations of Youth Tourism

> Ibiza, Ayia Napa, Sodom and Gomorrah: they are mere monasteries, we are told, compared to the organized bonk and boozathon that is Faliraki, where drunken damsels from the home counties are said to roll in the street and beg for sex from men rampaging in togas. (interview with Gerard 2003)

The summer of 2003 saw an elevated interest in the behaviour of British tourists abroad, both in the United Kingdom and in the tourist receiving countries of the Mediterranean. The above quote from a *Sunday Times* article is illustrative of the tone and content of much domestic British coverage of violence on the streets of Ayia Napa, the hedonistic behaviour of British tourists in Faliraki and the apparent moral decline of the nation. The most remarkable story of that summer was that of the Club 18–30 representatives alleged to have staged a live sex-show on a beach in Corfu, Greece – a story that led me to wonder why I was located in Faliraki that summer when the action was clearly elsewhere! Stories such as these provoke outrage and horror among elements of the British press and confirm the worst fears of those that imagine Club 18–30 and youth holidays more generally to be orchestrated and alcohol-fuelled holidays with sex as a part of the product. For others, the events on the beach in Corfu and the video recording said to exist of them have become an element of the mythology of the youth holiday. While it is doubtful that any surviving video of club representatives engaged in sex acts on Greek beaches is in wide circulation, it is certainly the case that at least one pornographic film has been inspired by the episode. The film *Filthy Club Reps* (2002) draws on a fantasy in which the holiday company employee exists to provide direct sexual gratification. Indeed, the very existence of such a film realizes some of the expectation that emerged from the popular ITV series *Club Reps*, not least in the sense that two of the stars of this low-budget pornographic film had first come to public attention as a part of that television series. TV series such as *Club Reps* and *Fantasy Island* (recorded in Ayia Napa) help to glamorize at the same time as normalizing aspects of typical British youth tourist behaviour in the Mediterranean. Such representations work to close off a symbolizing loop within which those previously having such experiences can reminisce while preparing for their next such holiday, those seeking such gratification can gain an insight or heighten their desire, and those who may be watching merely for entertainment or voyeuristic purposes may find themselves suddenly enriched by a host of new desires.

A browse along the shelves of second hand record shops, or CD stalls at car boot sales, will reveal to the casual consumer the prevalence of compilation CDs incorporating the words 'Ayia Napa' into their titles during the early years of the twenty first century. A more long-standing tradition is the even greater volume of similar CDs and DVDs concerned with clubbing holidays to Ibiza. These artefacts function alongside the television programmes and the specialist dance music press to motivate visits and to prepare the visitor for their holiday. The CDs and magazines orient attendance at particular clubs or venues, familiarize people with the DJs and tunes they can expect to encounter, and generally work to reinforce a sense of participation in a vibrant and desirable popular cultural form. All of these opportunities to consume aspects of a Mediterranean clubbing holiday before and after the event itself strengthen both the motivation and satisfaction of the consumer, and provide them with cultural capital with which to attempt to present themselves as knowledgeable insiders before, during and after the holiday itself.

Popular Cultural Tourism as Serious Leisure

As the above has shown, the embodied experiences of clubbing holidays do not simply emerge afresh for a fortnight each summer, but are inherently related to everyday practices at home. Clubbing holidays, including those to Ayia Napa, reflect the sorts of practices that Stebbins (1996, 2001) is referring to when he writes about enthusiasts participating in 'Serious Leisure'. Each year, large numbers of newspaper and magazine features, cover mounted compact discs, television shows and a glut of themed compilation albums create particularized notions of Ayia Napa as a holiday destination, a haven for clubbers and a hedonistic retreat, as well as reinforcing the strong association with garage music. The knowledges tied up in artefacts that represent Ayia Napa are crucial to a critical understanding of tourism in the resort. The constant stream of material about Ayia Napa has enabled the place to stand as a powerful metonym for the clubbing or youth holiday, encouraging more and more people to visit the town. Such banality is only achieved through constant restatement and reassertion such that the activity is entirely commonplace, something which often fails in the tourist context within which the notion of gazing on difference (Urry 1991) remains an important motivating factor. Stebbins' (1992) serious leisure thesis and ideas concerning the consumption of the familiar (Prentice and Andersen 2003) suggest that tourist consumption and behaviour should not be separated out from more general life experiences and everyday practice. Serious leisure is about the processes of identity creation and an extension of general leisure patterns indulged in at home, such that the holiday becomes not so much a break in activity as a short period of increased activity. Tourist behaviour is seen by Stebbins (1996) as a 'career-like' pursuit in that it is lifelong and is about the collecting of experiences as well as, more importantly, being about the (re)performance of identity and the construction of a biography. Thus, helpful to us here is the notion of the banal in that it expresses all of the flow of everyday and commonplace activities and occurrences that are neither highly dramatic nor unusual (Billig 1995, Palmer 1998).

The performance of public rituals, and the symbolic work involved in such performances, identifies both performers and spectators as either insiders or outsiders of a particular culture, society or milieu. Analyses of this kind have thus far in tourist studies been limited to considerations of middle-class, highly-educated and relatively affluent cultural tourists. This analysis seeks to incorporate some of the insights from these previous studies in an exploration of working class mass tourist cultures. A note of caution, however, is important here as UK Garage and dance music scenes more generally are largely independent of class identities, but dominated by working class youths whether self-identified or otherwise. The accumulation of cultural capital should be recognized as contingent and variable according to the characteristics of the persons seeking such wealth and those of their social groupings. As discussed above, the participants in youth holidays are gathering valuable experiences that have an exchange value within their friendship and wider peer groups, they are participating in a career-like pursuit and, in some cases, have a deeply-felt and dearly-held attachment to a subcultural scene and the practices surrounding it. These are the same kinds of attributions of affection that have been proposed as an explanation for more obviously elitist forms of cultural tourism, demonstrating that youth clubbing holidays are as *serious* as any other form of tourism. Two weeks of taking part in activities and consuming commodities familiar from home thus represents a speeding up of signification rather than the dramatic exotic break that tourist studies has historically concerned itself with.

Conclusions: The Unremarkable Holiday

The above discussion has demonstrated that there are many ways in which a youth holiday with an important clubbing element is simply like a fortnight of Friday and Saturday nights at home. It is not dramatically different, but it could be seen to be a spectacular manifestation of the everyday or every week banal in which people engage at home with often the same friends, in the same sorts of venues, listening to the same kinds of music, consuming the same sorts of drugs and having sex with very similar types of people. The representations consumed in advance heighten the sense of expectation and also the sense that such expectation has been realized during the vacation. The more spectacular manifestation simply reasserts the banal, reasserting the importance of the leisure pursuits of home for the serious tourist. While it is doubtful that serious is necessarily the right word, it is clearly the case that the youth tourist could be described as committed and as a cultural insider, accruing cultural capital for public display among other members of dance music communities and among other groups of youths with similar interests. It is likely that the re-siting in Greece or Cyprus is less important than the recitation of the rhetoric of UK Garage music through the set of practices that attach to the subculture. Ultimately, while there is much work still to be done on youth cultures, we should not be surprised or shocked that people like to go on holiday and do the things that they like to do at home. Equally, we should not be surprised that British youths behave in the Mediterranean in much the same way as they do on UK High Streets each weekend.

References

Anderson, B. 1983. *Imagined Communities*. Verso: London.

Bærenholdt, J.O., Hadrup M., Larsen J., and Urry, J. 2004. *Performing Tourist Places*. Aldershot: Ashgate.

Baudrillard, J. 1990. *Fatal Strategies*. New York: Semiotext(e).

Billig, M. 1995. *Banal Nationalism*. London: Sage.

Clift, S and Carter, S. 2000. *Tourism and Sex: Culture, Commerce and Coercion*. Continuum: London.

Clift, S. and Forrest, S. 1999. Gay men and tourism: destinations and holiday motivations. *Tourism Management*, 20(5), 615–25.

Coleman, S. and Crang, M. 2001. *Tourism: Between Place and Performance*. Berghahn: London.

Crouch, D. 1999. *Leisure/Tourism Geographies: Practices and Geographical Knowledges*. Routledge: London.

Debord, G. 1967. *The Society of the Spectacle*. London. Zone Books.

Franklin, A. and Crang, M. 2001. The trouble with tourism and travel theory. *Tourist Studies*, 1(1), 5–23.

Hannam, K. Sheller, M. and Urry, J. 2006. Editorial: Mobilities, immobilities and moorings. *Mobilities*, 1(1), 1–32

Knox, D. 2006. The sacralised landscape of Glencoe: From massacre to mass tourism, and back again. *International Journal of Tourism Research*, 8, 185–197.

Knox, D. 2008. Spectacular tradition: Scottish folk song and authenticity *Annals of Tourism Research*, 35(1), 255–73.

Latour, B. 1994. *We Have Never Been Modern*. Harvard: Harvard University Press.

Malbon, B. 1999. *Clubbing: Dancing, Ecstasy and Vitality*. London: Routledge.

McKee, A. 1997. The aboriginal version of Ken Done: banal Aboriginal identities in Australia. *Cultural Studies*, 11(2), 191–206.

Mordue, T. 2005. Tourism, performance and social exclusion in 'Olde York'. *Annals of Tourism Research*, 32(1), 179–98.

Nash, C. 2000. Performativity in practice: some recent work in cultural geography. *Progress in Human Geography*, 24(4), 653–64.

Opperman, M. 1999. Sex tourism. *Annals of Tourism Research*, 26(2), 251–66.

Palmer, C. 1998. From theory to practice: experiencing the nation in everyday Life. *Journal of Material Culture*, 3(2), 175–99.

Prentice, R. and Andersen, V. 2007. Interpreting heritage essentialisms: familiarity and felt history. *Tourism Management*, 28(3), 661–76.

Ryan, C. and Hall, C. 2001. *Sex Tourism: Marginal People and Liminalities*. London: Routledge.

Salandha, A. 2002. Music tourism and factions of bodies in Goa. *Tourist Studies*, 2(1), 43–62.

Sellars, A. 1998. The influence of dance music on the UK youth tourism market. *Tourism Management*, 19(6), 611–15.

Skelton, T., Valentine, G. and Chambers, D. 1998. *Cool Places: Geographies of Youth Cultures*. London: Routledge.

Stebbins, R.A. 1996. Cultural tourism as serious leisure. *Annals of Tourism Research*, 23(4), 948–50.

Stebbins, R.A. 2001. Serious leisure. *Society*, 38, 53–7.

Urry, J. 1991. *The Tourist Gaze*. London: Sage.

Chapter 9

Corrupted Seas: The Mediterranean in the Age of Mass Mobility

Pau Obrador Pons, Mike Crang and Penny Travlou

As we noted in the introduction we have chosen to produce a book about *Mediterranean* mass tourism. Not just mass tourism that happens to take place in the Mediterranean, but the Mediterranean variety and inflection of mass tourism. With this choice we wanted to emphasize the fact that (even) mass tourism has histories and geographies. Moreover, the space of tourism has often been methodologically fragmented and marginalized as a side effect of two trends in research. First, research on tourism practices (as in many chapters in this volume) has often proceeded empirically through case studies of single destinations. Partly, this may be down to the logistics and funding of research, partly the difficulties of comparative work. Second, the space of tourism has been marginal both in academia conceptually but also on the ground – where resorts are often at the end of the line, and at the edge of territories. Thus recent assessments on regeneration and decline in British resorts point to transport and communication links leaving them marginalized (Communities and Local Government Committee 2007), while for many traditions of area studies tourism areas are equally on the edge of the territory and the fading edge of culture. This scale and focus is a variant of a methodological nationalism that is by no means confined to work on tourism. Indeed, in historiography it has often seemed that history only occurs at the scale of the nation state (Bentley 1999).

However, we also want to outline some cautions on the imagining of the Mediterranean. The anthropology of Mediterranean studies has all to often given a homogeneity to the region, indeed we might say defined the region through key attributes such as cultures of 'honour' and 'shame' (Albera 2006) and notions of cultural survivalism, where the antique survives into modernity (Mitchell 2002). Critics have argued with some force and justification that while Braudel and others have argued for the Mediterranean as 'an ecological unit', anthropology has seen it as a culture area characterized by the presence of codes of honour and shame in gender relations of a hierarchical nature and in so doing ended up opposing the primitivism of the Mediterranean with the modernity of Europe (see examples in Albera 2006: 116). As such a regionalist anthropology has colluded with a literary trope that portrays the Mediterranean through a limited range of generally cultural stereotypes of its people (Shore 1995) and a *geohistoire* has spoken to a region grounded in climate and agriculture – which themselves ascribe a different temporality to the region. Indeed geography has long learnt to be wary of models of cultural areas that all too often suppress heterogeneity, internal conflicts between subcultures and tend to be founded on models of rustic society (Crang 1998: 21). Where tourism is addressed at all it is as a problem, for people, places and research, not as one of the engines forging a Mediterranean regional identity. So we begin by asking what it is that a Mediterranean focus offers, first in terms of destabilizing the usual categories of nation and place by focusing upon a maritime imaginary like Homer's wine dark sea, second by looking at the analytic risks and commercial uses of fixing and exoticizing the Mediterranean, and then, third, asking how mass tourism is refracted through Mediterranean practices and imaginations and the diversity of outcomes now emerging creating many Mediterraneans and many mass tourisms.

Thinking Fluidity through the Wine Dark Sea

There have been attempts to refocus that national vision, using oceanic links and flows as ways of rescaling our account and thinking of the emergence of relational capacities in places. Thus we have accounts of the maritime communities of the eighteenth century Atlantic circulating politics and radical ideas (Linebaugh and Rediker 2000) and the gross circulation of people, commodities and cultures (Gilroy 1993). Indeed we can find a counter tradition that points precisely to the maritime spaces of the globe and networks of linkage – often around trade be they the Indian Ocean trades or the colonial Pacific. If we look hard enough we can even find Hegel proclaiming it the 'axis of world history... We cannot conceive of the historical process without the central and unifying element of the sea' (1975: 171–2 cited in Cooke 1999: 291). The use of ocean links and flows is not limited to academic writing; it also pervades popular accounts of the Mediterranean, from poetry to television to popular music. Thus Catalan singers such as Joan Manuel Serrat, Lluís Llach and Maria del Mar Bonet have made frequent use of these metaphors to articulate a relational vision of the Mediterranean as a crossroads of cultures, peoples and civilizations. One of the most accomplished examples is Lluís Llach's *Un Pont de Mar Blava* (A Bridge of Blue Sea) a utopian vision of the Mediterranean that sketches a shared history told as a respect for difference and the cross fertilization of cultures.

This sort of fluid history then is the invocation not so much, or only, of a specific geographical space around the Mediterranean sea itself but of the much broader sense given by David Abulafia of a middle sea as a condition and way of relating between places, where parallels have been called the Euro-Arab Mediterranean, American Mediterranean (Gulf of Mexico), Northern Mediterranean (Baltic Sea) and Japanese Mediterranean (Abulafia 2005). The Mediterranean becomes template for seeing the world of networks, relations and flows through trade and exchange. In that sense our choice of a maritime centred region echoes understandings that 'places should be seen as like ships that can move around within various human and

material webs So places can be 'dislocated' since images, thoughts, photographs
or memories are on the move, by means of technologies and the moving flows of
people' (Mavrič and Urry 2009: 647). We appeal to a vision of the fluid centred
geography to highlight the unstable fabrication of a tourist space. Our centring on
a 'thalassological' region (Horden and Purcell 2006) emphasizes fluidity, flows
and transformation rather than fixed identities and fixed destinations. We might
not go so far as Hegel and say that the sea 'gives us an impression of limitlessness
and infinity, and when man feels himself part of this infinity, he is emboldened to
step beyond his narrow existence ... Land ... binds man to the soil; consequently
a whole series of ties attaches him to the locality he lives in' nor to dichotomize
as he does 'the stability of the inland regions and the roving character of coastal
life with all its contingencies' (1975: 160, 162 cited in Cooke 1999: 295). But the
focus on the shifting ocean does seem to open out a sense of the experimental, the
contingent, the transient, the mixed and changing – that also to a greater or lesser
extent clearly mark the socialities and experiences of tourism.

The Uses of the Mediterranean

Set against this fluid imaginary is the tendency to think of the Mediterranean
precisely as a space apart – and indeed a time apart. This essentializing view of
it as a unit, marked by specific cultures, temperaments and habits, comes through
in much of Braudel's work and certainly that of his followers. Braudel, for all his
brilliance carves out the 'true Mediterranean', bounded by blocks of territory like
the Sahara or opposing the Atlantic so that he focuses upon not fluidity but the:

> local, permanent, unchanging and much repeated features which are the
> 'constants' of Mediterranean history.... All Western writers who have at some
> time in their lives encountered the Mediterranean have been struck with its
> historical or rather timeless character. Like Audisio and Durrell, I believe that
> antiquity lives on round today's Mediterranean shores (Braudel 1972: 1239 cited
> in Cooke 1999: 291–2).

His true Mediterranean is precisely delimited from Spain below Catalonia,
France through Provence, across Tuscany to Trieste and down the Aegean, and
on the southern littoral extending to the palm grove line. Indeed Braudel links
the Mediterranean culture to both climate and soil where 'man [sic] gains his
daily bread by painful effort. Great tracts of land remain uncultivated and of
little use... the soil ... is responsible for the poverty in inflicts on people, with
its infertile limestone ... [and] the desert lies in wait for arable land' (Braudel
1995: 241–2).

In finding an essence to the region he is not alone but stands in dubious
company. The appeal to and the appeal of the climate as some unifying bond have
lurked at the heart of pleasure seekers, and designation of the Mediterranean as a

pleasure periphery to northern Europe. How else should we respond to the great traveller Richard Burton and his locating of 'sotadic zone' of sexual perversion he argues is 'geographical and climatic, not racial' and found in a region bounded westwards by the northern shores of the Mediterranean (N. Lat. 43 degrees) and by the southern (N. Lat. 30 degrees), with a depth of 780 to 800 miles including meridianal France, the Iberian peninsula, Italy and Greece, with the coast-region's of Africa from Morocco to Egypt. This forms a centre of gravity from which desires spread out threatening surrounding areas (Phillips 1999: 74). It is here he says

> the Vice is popular and endemic, held at the worst to be a mere peccadillo, whilst the races to the North and South of the limits here defined practise it only sporadically amid the opprobrium of their fellows who, as a rule, are physically incapable of performing the operation and look upon it with the liveliest disgust (Burton 1885, *The Arabian Nights, Terminal Note section D, 207* cited in Phillips 1999: 75)

Here is the flipside to the romantic celebration of the Mediterranean character – the regions positioning as ludic space, for visceral experience and social vices, which is still invoked in the marketing of destinations like Mykonos as gay resorts (Waitt and Markwell 2006: 73 and 118). If such then are the characterizations made of the Mediterranean then they point to the history of travel cultures as attracted by freedoms, by different and dangerous mores associated with the region. As such they point to not characteristics of the people or place but its creation and performance through repeated tropes and discourses.

Michael Herzfeld (2005) has long criticized the essentialisms of *Mediterraneanism* where a danger of circular argument, moves from *the Mediterranean* as a single and distinctive region or to its particular exemplifications in specific characteristics (typically cultures of honour, shame or machismo) and back again. Here we take rather his point that it is not some referential truth but the use of the term that makes the category valid and interesting (2005: 49–50), and its use is especially marked in tourism, to articulate a type of destination and type of holiday experience. The Mediterranean is performative in every sense. It is less about shared traits than the play of claims and knowledges about those shared traits. In that sense we are offering a rhapsodic Mediterranean that responds to and echoes the imaginations and desires expressed but a 'rhapsodic celebration is a celebration only of discourse about, or imagery of, the Mediterranean' (Horden 2005: 29). But it is more than that, for we do look to the commonalities articulated occasioned and created through that imaginary. As Inglis (2000: 124) notes, although artists like Cezanne popularized the scenery, it was a guidebook that first coined the phrase 'Côte d'Azur' in 1887, painting the landscape with a new brush in so doing.

So we do not see the 'timeless realities' as Braudel might but we do have to reckon with the 'Mediterranean as a unit, with its creative space' and connections through the sea, And for us this must be the 'greater Mediterranean' where Braudel speaks of its cultural filiations and trading connections. We must within the space of Mediterranean tourism include the origins of the tourists in predominantly northern Europe. It must also speak to the oppositions carved by *geohistoire* against the (*empty*) Sahara, Arabia and the Northern continent. There is then a risk of artificially accentuating the contrasts of say of Northern and Southern Europe (Albera 2006: 117). Moreover we must think of this reflexively to ask what work then the notion of an actual Mediterranean region does (Herzfeld 2005). In this context one avenue for us are its historical geographic connotations. First, is its position as defining both the border of a fortress Europe but also its expansive frontier and contact zone in the shared and turbulent history of Occident and Orient. Second, is its location as the origin of western civilization, and the shared history of the bearers of that civilization (principally the Romans) through classical legacies, ruins and their subsequent visitation.

First then, it serves politically to also point to the Mediterranean space as bridging the geopolitical divides of Europe and Africa, the West and the Near East. In this sense it dislocates a notion of a cultural divide say at the Straits of Gibraltar, reminding us that different historic visions of identities and regions have crossed this divide (Boer et al. 1995) and that if there was once a rationale for such a divide, then the flows of tourism suggest otherwise nowadays. In the words of the Multiplicity.lab the Mediterranean has become a solid sea of new proliferating connections and entanglements. There are the flows of migrant work, some to sustain and support tourist development that destabilize notions of traditional 'Mediterranean' national cultures with new migrant service classes in countries unused to immigration (Lazzaridis and Wickens 1999). So one can trace routes around the tourist industry, that lead in surprising directions, with circular flows of workers and cultures, artefacts and symbols across the sea (Jacquemet 1999). Indeed tourism is also one way to speak through the colonial and postcolonial entanglements of an industry that was once used to educate Europeans about colonial possessions, as part of a *mission civilisatrice* (Hazbun 2008: 5) indeed was a form of possession through touristic practice reducing the African littoral to the timeless and scenic to be pictured (Gregory 2001, 2003).

Alternately the myth of Mediterraneanism conjured by Albert Camus was designed to offer a bridge of shared history for colonizer and colonized by mobilizing the notion of shared region, '*la culture vivante*', openness to mixture and the classical legacy (Davison 2000). The hierarchy of values that the space of the Mediterranean conjures up (Herzfeld 2005) are then polysemous and steeped in contested histories – that speak through changing cultures of travel. Perhaps this comes together well in Tunisia where the undress of beach tourism and alcohol consumption were regarded as culturally inappropriate by many and 'mass beach tourism oriented to European desires proved an especially ripe target for Islamist tracts against westernizing and secularizing government policies and social trends towards consumerism and materialism' and in response the government sought to position Tunisia as a Mediterranean society at the crossroads of cultures – with a marketing campaign orchestrated around the Punic hero of Hannibal and the ancient city of Carthage (Hazbun 2008: 47, 70).

Second, then, the Mediterranean has commonalities through histories of tourism that are entangled in both defining difference from Europe but also shared cultural origins (Moulakis 2005). As Inglis (2000: 114 and 117) argues for Northern Europeans it has featured largely in the education of elites and been made special through the celebration of a special classical heritage and bond of imputed values. We do not need to go back to the grand tour to find these echoes. Witness Freud's own account of his visit to the Acropolis which outlines the contradictory geography and temporality of past and present, centre and margin for Greece (Crang and Travlou 2001). At a first level his response was shock at encountering the reality of something he had learnt from stories as a child, and had never expected actually to visit. Freud exclaimed: 'So all this really does exist, just as we learnt at school' (Gourgouris 1996: 122). The ruined Acropolis stood as a (material) metonym of the past (imagined, virtual) Hellenic civilization, or created, in Freudian terms, a derealization: 'signifying both the unreality of the experience in the light of the preconception and the evident unreality of the preconception in view of the experience' (Gourgouris 1996: 123). That gap of preconception and experience can be found writ large in the anxiety of Martin Heidegger who long delayed a voyage to actually visit the Greece with whose classical legacy he had spent a lifetime in dialogue, for fear the actual existing Greece might interrupt that conversation. Like Ruskin in Florence, he fretted that the bustle of tourists, their to-ing and fro-ing, and their very zeal to consume the classic sites 'threatened to degrade what was just now the element of our experience into an object ready-at-hand for the viewer' but even he did not dismiss the experience 'No one, however, would like to contest or underestimate the fact that several of them would preserve a serious impression from the temples of the Acropolis for the rest of their journey' (Heidegger 2005: 42 cited in Smith 2009: 613). This gap, or fear of a gap of preconception and experience runs in less rarefied veins as well. It is in part, the structure of tourism as a tension of desire, satiation, dissatisfaction and idealized memory. Papers in this collection have shown the magnetic appeal of the Mediterranean runs strongly through from the high brow, through the middle brow to the populist.

The classical heritage and values of the Mediterranean have featured strongly in southern as well as in northern Europe. There is a long tradition of cultural movements in the Mediterranean region that celebrate the order and beauty of classical culture and claim them in a history of cultural circulation around the middle sea. One such movement is *Nouncentisme* in Catalonia in the early 20th century that originated largely as a reaction against *Modernisme* (Resina 2008). This movement turns to classical culture to reposition Catalonia within the reflected glory of the cultural superiority Mediterranean City states afforded. This stands in contrast to *Modernisme*, which located Catalonia in relation to the liberal Northern European Countries. The movement – conceived as a journey rediscovering Classical roots – seeks to translate, literally and metaphorically, the principles of Greek and Roman culture, in particular the idea of civility associated with the Greek Polis, to an emergent country (Resina 2008). Among other things it articulates a new utopian vision of Barcelona, key elements of which are still with us today. The Barcelona model of urban regeneration (Monclús 2003) draws heavily on these antecedents in rejecting the myth of low density north European city in favour of the Mediterranean tradition of high density cities and public spaces – precisely the elements and associations that have made the city a tourist success.

There is something then of an exceptionalist claim in our stories of the Mediterranean (Horden and Purcell 2006) – though we would hope a modest and defensible one. The Mediterranean is not just any tourist destination; it is the first international one and, with more than 230 million tourists a year, the largest destination for mass tourism in the world. As Löfgren points out, 'if we want to understand the making of package tourism as a giant industry we should start on these shores. The models of tourism that emerged here were later exported to other parts of the world, as global tourism expanded' (Löfgren 1999). So focusing upon the Mediterranean allows us to speak to important and influential forms, while emphasizing the fact that not all mass tourism is the same. Despite the fast pace of globalization, it remains a largely regional phenomenon, with its regional and local idiosyncrasies. We should not forget that the vast majority of international tourism takes place within the same region – Europe, America or Asia – and that the growth of *regional* travel, within the same continent, has clearly outnumbered the growth in truly *global* travel, between continents.

As we have pointed out in the introduction, we feel uneasy about the way mass tourism has been conceptualized as a free floating phenomenon that is imposed on destinations and has nothing to do with the place where it occurs. Clearly there are elements of deterritorializing processes in mass tourism, for instance in the production of standard international style of hotel complex in the 50s and 60s with the rise of standardized beach tourism, and standardized architectures that become substitutable from place to place (Hazbun 2008: 8). It is at a regional level across the Mediterranean that they end up locked in a downward spiral of price competition due to oversupply. It is at the Mediterrananean level there is an emergent form of mass tourism shared across these locations. Recent trends to reterritorialize tourism, by linking it, however instrumentally, to local cultures, creates an overall Mediterranean symbolic economy framing these localisms and expectations. By locating mass tourism in the Mediterranean we wanted to make an exercise of situated thinking, grounding tourist activity within a social and geographical society. Mass tourism is all about place – although not in the narrow organic sense.

Mass Tourism in the Med

We thus conclude this book with a commentary on the Mediterranean in the age of mass tourism. By focusing on the *Med*, this book has emphasized the significance of mass tourism in shaping and reshaping the region. Mass tourism has become the very fabric of the Mediterranean, conferring to the region a new economic and social centrality. The movement of tourists and those who make their living from them is what today holds together the old Mare Nostrum. Clearly then we are responding to Fernand Braudel, whose path breaking study of the Mediterranean world in the age of Philip II combined like no one before economic and geographical elements, material, social and symbolic dimensions to produce what he called a total history. The actors of his work are not kings and generals, but the sea, the winds, the trader and the sailor. He focuses on the unspectacular, the recurrent patterns of life, and the everyday existence in the Mediterranean, seeking 'those local, permanent, unchanging and much repeated features which are the constants of the Mediterranean history' (1995: 658). Without sharing his totalizing aim, the chapters of this book have also tried to reconstruct the complexities of the Mediterranean geographies and sensibilities by focusing on the unspectacular, the recurrent patterns of life. But whereas Braudel focused on the dawn of modernity in the days of Philip II, this book has looked at the formation of the post-modern Mediterranean. Reconstructing tourist skills and sensibilities, this book is a (very modest) contribution to an archaeology of the Mediterranean present.

Whatever the changes over the last 400 years, Braudel's characterization of the Mediterranean as a space unified by movement still holds true. 'The Mediterranean' – Braudel explains – 'has no unity but that created by the movement of men, the relationships they imply, and the routes they follow' (1995: 201). There are many different mobilities shaping the Mediterranean present, the most important of which have to do with tourism: the charter planes linking Frankfurt and Palma, Lanarca and London Gatwick; the holiday cruise full of American tourists drifting from Barcelona to Marseille, Genoa, Naples, Tunis, Piraeus and finally Istanbul; the seasonal migration of workers from the village to the coast and from the southern shore to the northern shore; and the dreams of happiness pushing north Europeans to travel south every summer and which at the end of the holiday mutate into resentment and frustrations on the way back north. Today the flows of tourism rather than trade or agriculture constitute the very fabric of the region and create tourist places:

> Tourist places are hybrid systems of material and mobile objects, technologies and social relations that are produced, embodied, imagined, memorized and anticipated. Moreover, places are not fixed locations on a map but are better viewed like ships, moving around the globe; they travel around as images and memories and move closer to global centers or further away (Mavrič and Urry 2009: 650).

If the immemorial routines of olives, wine and grain in Braudel's characterization have declined then there is instead a new and no less seasonal rhythm to Mediterranean life. The Mediterranean has its own characteristic rhythms. However what defines the calendar is not anymore the grape harvest of the late summer as the arrival of the first charter plane in May. At the heart of it all, there is the seasonality of the climate, which structures the tourist season in a similar way that it structures the agricultural year. The economies of sun, sea and sand are heavily dependent on very particular climatic conditions, which are predominant in July and August. The optimal conditions for mass tourism include absence of rain and wind, maximum day time temperature around 30°C and the optimal 24 hour mean temperature around 24°C (Amelund and Viner 2006). The sun continues to shape the economic and social life of the Mediterranean. The rhythms and materials of mass tourism give coherence to the Mediterranean present and that is producing a sense of déjà vu, with Braudel's observation that: 'In the sixteenth century, a native of the Mediterranean, wherever he might come from, would never feel out of place in any part of the sea' (1995:178). Despite the deep political, cultural and religious divisions, the same is true today. Tourism cultures are fuelling these commonalities building on shared and contested histories so that Essaouira in Morocco so closely echoes French colonial styles that Place Moulay El Hassan, can be written up, in a travel guide as a large square fringed with bars and cafes serving Italian ice cream and cappuccinos, 'that would not be out of place in southern Europe' (Guardian 5 June 2004). Wherever we go in Spain, in Tunisia, in Cyprus or in Greece, in Turkey, we will find to the point of obsession the same hotels, the same golf courses, the same marinas, the same low quality foods, the same disputes for sun beds, the same 1.5 litre plastic bottles of still water buried in the sand, the same smell of sun cream.

The Mediterranean and its harbours lost a few centuries ago the privileged commercial position that enjoyed at the dawn of modernity. With the emergence the northern European powers, the consolidation of the world trade routes and the decline of medieval city states, the Mediterranean ceased to be the axes of western civilization and economy to become its southern periphery. Such a marginal and peripheral identity still heavily conditions the Mediterranean present. The 'Mare Nostrum' of the Romans is today a major dividing line in the world, separating Europe and Africa and Asia, the Christian, the Muslim and the Jewish world, the first and the third world that produce problems of governance. It is precisely these problems that attract the attention of researchers with an interest in the Mediterranean. As Gillespie and Martin point out 'Researchers more commonly are drawn into learning about the area by the issues that are presented as 'problems' in need of policy solutions' (2006: 152). Mediterranean studies are dominated by the security-driven agenda of international relations. At the centre of this agenda there is the Euro-Mediterranean Partnership or Barcelona Process, which started in 1995 with the Barcelona Euro-Mediterranean Conference. This process is a framework of bilateral and multilateral relations leaded by the European Union that promotes peace stability, cooperation and free trade in the newly reminted and recast 'Mediterranean region' (Liotta 2005).

Rather than challenging this peripheral identity of the Mediterranean, the dominant perspectives on mass tourism have tended to reinforce its marginality. Mass tourism has been conceived as an agent of domination that transforms less developed regions into peripheries of the industrialized richer areas. At its starkest Frantz Fanon turned the 'industry into a metaphor for the brothelization of the Third World at the hands of leisure imperialists' (Hazbun 2008: 5). This line of thought imbues for example the influential work of Turner and Ash who developed the notion of pleasure periphery back in the 1970s. 'Today it is the Nomads of Affluence, coming from the new Constantinoples – cities like New York, London, Hamburg or Tokyo – who are creating a newly dependent, social and geographical realm: The Pleasure Periphery' (1975: 1). The Mediterranean is one, if not the most important one of these pleasure peripheries which are geographically conceived 'as the tourist belt which surrounds the great industrialized zones of the world. Normally it lies some two to four hours' flying distance from the big urban centres, sometimes to the west and east, but generally toward the equator and the sun' (1975: 1–2). In this scheme of thought mass tourists are described as 'the Barbarians of our Age of Leisure, the Golden Hordes swamping less dynamic societies'. Similar views can be found elsewhere.

Turner and Ash's view of the Mediterranean is not very different from the perspective of international relations. In both cases the Mediterranean is conceived as a conflictual region at the edge of Europe, swamped either by civil wars or by the barbarism of tourists, a region that has no logic of its own. The defining element of the region is the relation of dependency with richer and more powerful areas of northern Europe. The European Union policy language of 'lagging regions' has reinforced the sense of social temporality and discourses of modernization that has exoticized the Mediterranean. This metaphor of dependency has proved very popular in Mediterranean itself not least because it justifies European investment and exempts tourist regions from their responsibilities for the situation. It is always easier to blame foreigners.

We remain unconvinced about the adequacy of metaphors of marginality and dependency in the case of mass Mediterranean tourism. Instead of explaining its functioning, these metaphors reinforce the peripheral identity of the Mediterranean, establishing a hierarchical value system that denies the modernizing capacity of mass tourism. Romanticizing a bygone era dominated by organic communities, the main focus of these metaphors is on what has been lost. There is no value placed on the modern Mediterranean, unless it remains fixed in a pre-modern era. Conveying a rather paternalistic attitude, the notion of dependency attributes a passive and secondary role to the Mediterranean. It also obfuscates precisely the production of that timeless and static sense of the past. Thus, for instance, it omits the folkloric definitions of rustic types in setting apart and framing the contemporary Greek peasantry for the viewer as exemplars of Mediterranean types (Tzanelli 2003), and indeed the creation of timeless exotic types for the metropolitan north framed in exhibitions such as *L'Exposition Coloniale Internationale de Paris* (1931) where Marechal Lyautey created almost a suburb of Paris which restaged the colonial policy of separating modern metropolitan styles from 'native' styles for the pavilions of the colonies, exemplified by forms like Tunisian souks (Çelik and Kinney 1990, Morton 1998). In that separation it mirrored the policy on the ground of Lyautey during his service as governor of Morocco, where his stated aim was:

> to conserve in Morocco Beauty – [...] Nothing has been more deadly to the originality and the charm of Algerian cities, of so many oriental cities, that their penetration by modern European installations [...] The preservation of the native towns is not only a question of aesthetic satisfaction [...] but a duty of the state. Since the development [...] of tourism on a large scale, the preservation of the beauty of the country has taken on an economic interest of the first order (Lyautey 1927 cited in Minca 2006: 162).

Lyautey's *politique indigène* intended that 'indigenous peoples' should 'evolve in autonomous fashion', by 'keeping everyone in his proper place, appropriate to his role in society' (Rabinow 1989: 28). It built new 'European' cities alongside preserved and fixed 'indigenous' Medinas (Çelik 1997).

The timeless cultures of the Mediterranean then have been reinvented through the actions of various travellers and representations, and almost always under the sign of their own disappearance, as when Edith Wharton recounts her arrival in Lyautey's Marrakech:

> the strange survival of mediaeval life, of a life contemporary with the crusaders, with Saladin, even with the great days of the Caliphate of Bagdad, which now greets the astonished traveller, [but which] will gradually disappear [... since despite] the incessant efforts of the present French administration to preserve the old monuments of Morocco from injury, and her native arts and industries from the corruption of European bad taste, the impression of mystery and remoteness which the country now produces must inevitably vanish with the approach of the 'Circular Ticket' (Wharton 1920: ix-x).

Here then we see the production of the timeless, its celebration and also a fear that a new world of mass mobility will de-distantiate the Mediterranean making it too close, too accessible and too frankly ordinary. Remarkably she echoes the fears of Bensusan's elegiac travelogue from twenty years earlier which feared that as 'a result of French pacific penetration, flying railway trains loaded with tourists, guide-book in hand and camera at the ready, will pierce the secret places of the land' (Bensusan 1904: 77). The corrupting presence of tourism is a continual theme in narratives that see the Mediterranean as comprising fixed non-modern cultures that are to be eroded by mass tourism. Yet tourism that continues to trade upon those very identities and essentialized images when *Conde Naste* travel magazine in 2005 still celebrated a Marrakech where 'Time has come to a standstill in this enchanting city' and the city presents 'the chaotic surface of noisy automobiles, mopeds, donkey carts' and a people outside of modernity 'living their days much as their ancestors did'.

Mass tourism challenges the current marginality of the Mediterranean within the world stage through creating new cultures and mass cultural phenomena. The millions of *Golden Hordes* that invade every summer the Mediterranean beaches have given a new economic and social centrality to the region. With all its social unbalances and negative environmental effects, mass tourism has demonstrated itself to be a formidable engine for economic development and cultural change, transforming derelict economies. Instead of seeing it as a past world corrupted by hordes we see new social and economic formations embracing North and South. There are elements of Fordist mass leisure, alongside atavistic reimaginings of petty capitalism, niche marketing and flexible specialization (Torres 2002) and these developments are occurring not against traditional economies or against mass tourism but as part of densely variegated landscape of new social forms created through tourism.

Contrary to what most people think, this process has been driven from the south as much as from the north. There is a need to de-romanticize the academic view on the Mediterranean and mass tourism and stop thinking this region in melancholic – touristic – terms as yet another example of a lost paradise invaded by flocks of tourists. We need to take seriously what is banal, fun and depthless, we need to examine the production of new bodies, their capacities and knowledges. What we see here is not an invasion or a loss but the transformation of play and leisure into one of the strongest forces in our culture, and its interplay with geographical identity. It is time to put mass tourism and the Mediterranean in relation to European modernities and post-modernities as part of the new economic and social modes of organization. The hypermodern and the antique in the Mediterranean jostle each other cheek by jowl (Leontidou 1993). Here then there will be the revalorization of tradition, and its political and commercial contestation and appropriation (Herzfeld 1991, Williams and Papamichael 1995). There will be the performance of the fixed and timeless imaginary of the Mediterranean. There will be the immobilization of some people, and the mobilization of others – rich and poor.

Many Tourisms, Many Mediterraneans

As well as conferring a new centrality to the Mediterranean, mass tourism has introduced new dimensions to its geographical identity, complicating the picture of the region with yet another set of layers. Mass tourism has introduced a logic excess, hedonism and luxury, as well as an appreciation for culture, history and heritage. With an increasing number of people experiencing the Mediterranean through practices of consumption and leisure, these logics have become closely associated with the region. However these new logics have not replaced the old ones, The coexistence of multiple logics often unrelated to each other is a defining feature of the Mediterranean present. A political logic of division and conflict coexists with the bio-political logic of play, fun and entertainment that does not understand the reasons of conflict. The Mediterranean of those who dwell permanently as residents sits side by side with the Mediterranean of fleeting tourists dwelling-in-mobilities. The striated space of the thwarted and obstructed movements of migrants, is the same as the smoothed flows of tourists. The indigene, the returnee and the resident tourist may all seek traditional styles of housing, the traditional *riad* (courtyard house) may be preserved because 40 percent of the Medina of Marrakech becomes foreign owned. The Mediterranean in the age of mass tourism is an elusive region marked by contradiction and complexity. It is simultaneously made of play and work, fun and grief, conflict and feast, mobility and immobility, big yachts and the small *pateras*, tourists and illegal migrants. It does not allow a one-dimensional reductive view. In sharp contrast with the dominant image of the Med, which tends to emphasize a single facet of the region, this book has tried to reveal a kaleidoscopic picture with multiple angles and dimensions. Capturing its complexity and its fluidity is the main challenge of any book about the Mediterranean.

Mass tourism has not created a homogenous region. Pleasures have differed and in general social classes have drawn apart as 'capitalism orders these divisions of labour, of pleasure and of pocket' (Inglis 2000: 110). There are many tensions underpinning Mediterranean tourism, the most significant of which is perhaps between culture and hedonism. On the one hand Mediterranean tourism is educative, aesthetic, solitary and formal. A number of scholars in this book explore this dimension, including Haldrup, Crang and Travlou. Exploring the birthplace of the old civilizations and engaging with the origins of western culture have been major driving forces of tourism development. The Mediterranean is the place of Ancient Rome, the Greek temples, the cities of the Italian Renaissance and the Egypt of the Pharaohs, all of which pull together extraordinary interest. As Inglis points out 'Our Mediterranean associations take us rather nearer to the mysterious origins of civilization than one would expect from the genial lubricity of Club Med' (2000: 116). The origin of this form of tourism, which Urbain (2003: ix) identifies with the character of Phileas Fogg, has to be found in the grand tour. This journey was meant to finish the education of the upper classes of Britain in the seventeenth and eighteenth centuries with a first hand knowledge of classical culture, renaissance and the exotic otherness of the south. However there is another face of Mediterranean tourism, which is increasingly prominent today but it is not something new. Mediterranean tourism is also gregarious, hedonistic, liberated and unhistorical. Enjoying, having fun and getting away from it all have also been a driving force of tourism, even in the time of the grand tour. This second form of tourism corresponds with the fictional character of Robinson Crusoe (Löfgren 1999: 9, Urbain 2003: 14–20) and is evident in the chapters of Pau Obrador, Karen O'Reilly and Javier Caletrío.

Intimately linked with this duality, there is the tension between the familiar and the exotic dimensions of the Mediterranean. For some the Mediterranean is a world of habits and routines, a place of dwelling and return. This *anti-touristic* dimension is visible in the villas of Dènia, the family hotels in Menorca and the retirement apartments in Torremolinos, which offer a promise of simplicity and happiness and the recreation of familiar landscapes with its social networks. For many other people, including the Danish tourists Haldrup describes in his chapter, the Mediterranean continues to be an exotic destination full of magic wonders, mysterious cultures and lost civilizations, a place that even functions as a gateway to the east. The Mediterranean of mass tourism emerges as a place of contradiction, a place of culture, but also of swimming and sunbathing; it is a place to discover the origins of civilization but also for relaxing from the constraints of urban stress; it is today still an educative destination but one where that learning encompasses the realms of romance and sexual adventure.

We do not know what the future of the Mediterranean holds, whether mass tourism will continue to shape so intensively the Mediterranean way of life. At the moment, there is no sign pointing to the decline of mass tourism, indeed all evidence suggests its continuous expansion. For sure there will be changes and fluctuations so there will not be one simple model of mass tourism, but the fusion of forms – from neotribal youth cultures, to democratized cultural tourism, to seasonal and retirement migrations, alongside the forms of provision for these from new *urbanaciones*, leisure facilities, local traders and large corporations. The plural landscape of touristic activities makes us conclude that whatever happens the Mediterranean will continue to be an elusive region marked by contradiction and complexity, in which multiple logics coexist.

There are also many uncertainties clouding mass Mediterranean tourism. First, least often voiced in the industry but perhaps most profound affects the fundamental and unchanging givens – the climate and environment. Climate change, and concern over it, might alter the currently dominant tourist sensibilities. A positive attitude towards sunlight, swimming and warm climates constitutes the basis of the Mediterranean tourism today. As Carter points out, 'the sun unproblematically condenses and signifies the essence of modern travel and sensuous pleasure'(2007:3). The predicted changed in the Mediterranean climate as well as the growing medical concerns about sun exposure introduce a high degree of uncertainty about these tourist sensibilities. We cannot avoid wondering about the reaction of mass tourism to changing climatic conditions that according to Amelund and Viner (2006) will make Mediterranean summers increasingly unbearable or about the effects on mass tourism of the changing social and medical attitudes towards the sun, which questions the positive association between health and sunlight. The dependency of mass Mediterranean tourism on such particular climatic conditions, so long taken for granted, and sensibilities about climate and travel is the perhaps its all too obvious Achilles heel.

Second, the generalization of a touristic stance in Western European societies constitutes another major uncertainty for Mediterranean tourism, which for a long time has relied upon a sense of the different and the unusual. The possibility of escaping the rigid constraints of everyday life in industrial societies has been a key attraction of Mediterranean holidays, in particular of more hedonistic destinations such as Eivissa and Mykonos. Today the contrast between touristic and non-tourist societies is blurring and the everyday life in the west is becoming increasingly indistinguishable from the touristic world. 'Tourism is no longer a specialist consumer product or a mode of consumption' explain Franklin and Crang (2001: 6). These processes of de-differentiation challenge a basic principle of Mediterranean tourism. What is the point of travelling to Mallorca for a stag do when the night life in Newcastle or Hamburg can be as good, or even better? What if the Mediterranean appeals for a weekend break or a golf break as much as a Northern city? What happens when the distance separating the holiday region and the home breaks down? A destination map of a low cost airlines offers a glimpse to the future. These companies are not just connecting northern European cities with the Mediterranean resort, as the charter companies used to do, Low cost airlines are connecting Europe in all directions – suggesting a multiplication of tourist destination and sensibilities. The list of uncertainties does not finish here. The desperately constraining temporality of the resort life-cycle starts to give way to plural histories and developments with these plural scales and logics. When demands are multiple and fragmented we may expect patterns of efflorescence and retrenchment in more complex sequences. We wonder as well about the possible clashes between the political and bio-political logics underpinning the Mediterranean present. The opposed nature of these logics were graphically evident the summer of 2007 when news bulletins showed images of exhausted illegal migrants arriving in small boats to the beaches of the Canary Islands at the peak of high season. To what extent can mass tourism continue to be oblivious to the huge political divisions of the Mediterranean? Can such contradictory logics continue to coexist so easily? And of course we wonder also about the excessive character of mass tourism. Like contemporary economic crises, the limits of mass tourism are not those of shortage but of excess, with excess of speculation, excess of hedonism, excess of drugs and alcohol, excess of pollution and water consumption.

References

Abulafia, D. 2005. Mediterraneans, in *Rethinking the Mediterranean*, edited by W.V. Harris. Oxford: Oxford University Press, 64–93.
Albera, D. 2006. Anthropology of the Mediterranean: between crisis and renewal. *History and Anthropology*, 17(2),109–33.
Amelung, B. and Viner, D. 2006. Mediterranean tourism: exploring the future with the tourism climatic index. *Journal of Sustainable Tourism*, 14(4), 349–66.
Bensusan, S.L. 1904. *Morocco: Painted by A.S. Forrest*. London: Adam and Charles Black.

Bentley, J.H. 1999. Sea and ocean basins as frameworks of historical analysis. *Geographical Review*, 89(2), 215–24.

Boer, P.D., and Bugge, O. 1995. *The History of the Idea of Europe*. London: Routledge.

Braudel, F. 1995. *The Mediterranean and the Mediterranean World in the Age of Philip II*. 2nd Revd. Edition. Berkeley: University of California Press.

Carter, S., 2007. *Rise and Shine: Sunlight, Technology and Health*. Oxford: Berg Publishers.

Çelik, Z. 1997. *Urban Forms and Colonial Confrontations: Algiers under French Rule*. Berkeley, CA: University of California Press.

Çelik, Z., and Kinney. L. 1990. Ethnography and exhibitionism at the Expositions Universelle. *Assemblage*, 13, 34–59.

Communities and Local Government Committee, Second Report of Session 2006–07, *Coastal Towns*, HC 351: HMSO

Cooke, M. 1999. Mediterranean thinking: from Netizen to Medizen. *Geographical Review*, 89(2), 290–300.

Crang, M. 1998. *Cultural Geography*. London: Routledge.

Crang, M. and Franklin, A. 2001. The trouble with tourism and travel theory. *Tourist Studies*, 1(1), 5–22.

Crang, M., and Travlou. P. 2001. City and topologies of memory. *Environment and Planning D: Society and Space*, 19, 161–77.

Davison, R. 2000. Mythologizing the Mediterranean: the case of Albert Camus. *Journal of Mediterranean Studies*, 10(1–2), 77–92.

Gillespie, R. and Martín, I. 2006. Researching the Mediterranean and the Middle East in the United Kingdom, Spain and Europe: present challenges and future initiatives, in *Researching the Mediterranean*, edited by R. Gillespie, and I. Martín. Madrid: British Council with IEMed and Fundació CIDOB, 151–75.

Gilroy, P. 1993. *The Black Atlantic: Modernity and Double Consciousness*. Cambridge MA: Harvard University Press.

Gourgouris, S. 1996. *Dream Nation: Enlightenment, Colonization and the Institution of Modern Greece*. Stanford: Stanford University Press.

Gregory, D. 2001. Colonial nostalgia and cultures of travel: spaces of constructed visibility in Egypt, in *Consuming Tradition, Manufacturing Heritage: Global Norms and Urban Forms in the Age of Tourism*, edited by N. AlSayyad. London: Routledge, 111–151.

Gregory, D. 2003. Emperors of the gaze: photographic practices and productions of space in Egypt, 1839–1914, in *Picturing Place: Photography and the Geographical Imagination*, edited by J. Ryan and J. Schwartz. London: I.B.Tauris, 195–225.

Hazbun, W. 2008. *Beaches, Ruins, Resorts: The Politics of Tourism in the Arab World*. Minneapolis: University of Minnesota Press.

Herzfeld, M. 1991. *A Place in History: Social and Monumental Time in a Cretan Town*. Princeton, N.J: Princeton University Press.

Herzfeld, M. 2005, Practical Mediterraneanism: excuses for everything, from epistemology to eating, in *Rethinking the Mediterranean*, edited by W.V. Harris, Oxford: Oxford University Press, 45–63.

Horden, P. 2005. Mediterranean excuses: Historical writing on the Mediterranean since Braudel. *History and Anthropology*, 16(1), 25–30.

Horden, P. and N. Purcell. 2006. The Mediterranean and 'the New Thalassology'. *American Historical Review*, (June), 722–40.

Inglis, F. 2000. *The Delicious History of the Holiday*. London: Routledge.

Jacquemet, M. 1999. From the Atlas to the Alps: chronicle of a Moroccan migration, in *Cities and Citizenship*, edited by J. Holston. Durham, NC: Duke University Press, 242–53.

Lazzaridis, G. and Wickens, E. 1999. 'Us' and the 'Others': ethnic minorities in Greece. *Annals of Tourism Research*, 26(3), 632–55.

Leontidou, L. 1993. Postmodernism and the city – Mediterranean versions. *Urban Studies*, 30(6), 949–965.

Linebaugh, P. and Rediker, M.B. 2000. *The Many-Headed Hydra: Sailors, Slaves, Commoners, and the Hidden History of the Revolutionary Atlantic*. London: Verso.

Liotta, P.H. 2005. Imagining Europe: symbolic geography and the future. *Mediterranean Quarterly*, 16(3), 67–85.

Löfgren, O. 1999. *On Holiday: a History of Vacationing*. Berkeley, CA: University of California Press.

Mavrič, M. and Urry, J. 2009. Tourism studies and the new mobilities paradigm (NMP), in *Handbook of Tourism Studies*, edited by T. Jamal and M. Robinson. London: Sage, 645–58.

Minca, C. 2006. Re-inventing the 'Square': postcolonial geographies and tourist narratives in Jamaa el Fna, Marrakech, in *Travels in Paradox: Remapping Tourism*, edited by C. Minca and T. Oakes. Lanham, MD: Rowman & Littlefield Publishers, 155–84.

Mitchell, J.P. 2002. Modernity and the Mediterranean. *Journal of Mediterranean Studies*, 12(1), 1–23.

Monclús, F.J. 2003. The Barcelona model: and an original formula? From reconstruction to strategic urban projects (1979–2004). *Planning Perspectives*, 18(4), 399–421.

Morton, P. A. 1998. National and colonial: The Musée des Colonies at the Colonial Exposition, Paris, 1931. *The Art Bulletin*, 80(2), 357–377.

Moulakis, A. 2005. The Mediterranean region: reality, delusion, or Euro-Mediterranean project? *Mediterranean Quarterly*, 16(2), 11–38.

Phillips, R. 1999. Writing travel and mapping sexuality: Richard Burton's Sotadic Zone, in *Writes of Passage: Reading Travel Writing*, edited by J. Duncan and D. Gregory. London: Routledge, 70–91.

Rabinow, P. 1989. *French Modern: Norms and Forms of the Social Environment*. Chicago, IL: Chicago University Press.

Resina, J.R. 2008. *Barcelona's Vocation of Modernity*, Stanford, CA: Stanford University Press.

Shore, C. 1995. Anthropology, literature, and the problem of Mediterranean identity. *Journal of Mediterranean Studies*, 5(1),1–13.

Smith, M. 2009. Ethical perspectives: exploring the ethical landscape of tourism, in *The Handbook of Tourism Studies*, edited by T. Jamal and M. Robinson. London: Sage, 613–30.

Torres, R. 2002. Cancun's tourism development from a Fordist spectrum of analysis. *Tourist Studies*, 2(1), 87–116.

Turner, L. and Ash, J. 1975. *The Golden Hordes: International Tourism and the Pleasure Periphery*. London: Constable.

Tzanelli, R. 2003. 'Casting' the Neohellenic 'Other': Tourism, the culture industry and contemporary Orientalism in 'Captain Corelli's Mandolin' (2001). *Journal of Consumer Culture*, 3(2), 217–44.

Urbain, J.D. 2003. *At the Beach*. Minneapolis: University of Minnesota Press.

Waitt, G. and Markwell. K. 2006. *Gay Tourism: Culture and Context*. New York: Haworth Hospitality Press.

Wharton, E. 1920. *In Morocco*. Illustrated Edition. New York: Charles Scribner and Sons.

Williams, W. and Papamichael. E.M. 1995. Tourism and tradition: local control versus outside interests in Greece, in *International Tourism: Identity and Change*, edited by M.F. Lanfant, J. Allcock and E. Bruner. London: Sage, 127–142.

Index